WILLIAM G. DAVIES

Assistant Professor of Chemistry
St. Mary's College
South Bend, Indiana

INTRODUCTION TO CHEMICAL THERMODYNAMICS:

A NON-CALCULUS APPROACH

1972   W. B. SAUNDERS COMPANY· PHILADELPHIA · LONDON · TORONTO

SAUNDERS GOLDEN SERIES

W. B. Saunders Company: West Washington Square
Philadelphia, Pa. 19105

12 Dyott Street
London, WC1A 1DB

1835 Yonge Street
Toronto 7, Canada

Introduction to Chemical Thermodynamics: A Non-Calculus Approach SBN 0-7216-2935-0

Print No.:   9   8   7   6   5   4   3   2   1

This book is affectionately dedicated

to Dr. A. H. Spong

of the University of Cape Town, from whom
I first learned that chemical reactions
don't sommer happen.

PREFACE

For several years now I have been convinced that it is a mistake to introduce students to chemical thermodynamics without first giving them some insight into what chemical equilibrium means on a molecular and statistical level. In my experience, students are only antagonized by thermodynamics if such an initial molecular approach is not given. Most students, particularly those with a chemical bent, have a very strongly rooted intuition that the position of a chemical equilibrium depends on the *nature of the molecules participating in it.* They quickly become suspicious of classical thermodynamics because it does nothing either to explain this intuition or to explain it away. Instead the subject is developed around such seeming irrelevancies as reversible heat changes or the efficiency of heat engines. Interest wanes and disenchantment grows as it becomes clear that molecules are not to be mentioned at all.

This book has been written in response to the difficulty just described. In place of the conventional macroscopic approach, the subject is treated here in microscopic and statistical terms right from the start. The student is never allowed to forget that *molecules* are involved in chemical equilibrium. Energy levels and the Boltzmann distribution are presented near the beginning of the book, and their relevance to chemical equilibrium is established. Only then are the first two thermodynamic functions, E and S, introduced.

By approaching the subject in this way, chemical equilibrium is not reached at the end of the book after a long and abstract argument involving subtle points in mechanical engineering. Instead it becomes concrete and *chemical* in character. There is no longer any need to fight against the student's instinctive prejudice in favor of molecules. On the contrary, he can be encouraged to look at simple reactions and predict whether they can proceed or not, by using purely *molecular* criteria.

I have deliberately written this book at a freshman level rather than for more advanced students. It seems to me that the purpose of a general chemistry course is to show students how chemists look at matter. Such a course is useful both to students who will be taking further chemistry courses and to those who will not. The more such a course is able to deal with the central insights of chemistry, the better it becomes. There can be no disputing the centrality of the insight which thermodynamics provides for chemists. If the subject can be made meaningful to freshmen, it should be included in the course.

There is a further advantage to be had from devoting some time to a molecular approach to thermodynamics in the freshman year. It enables classical thermodynamics to be covered much more quickly in later years. This opens up a little space in the notoriously overcrowded physical chemistry curriculum.

Because this book has been aimed specifically at freshmen, the ground that it covers is necessarily limited. Very little subject matter which is not directly related to chemical equilibrium is considered. The functions E, S, H, and G are introduced in that order, and the relationship between standard free energy changes and the equilibrium constant is established. No attempt is made to deal with subject matter such as the Phase Rule, the Gibbs-Helmholtz equation or Maxwell's relations.

The mathematical level of the book has been deliberately kept low. In particular, no calculus is required. Any mathematics beyond the level of high school algebra which is needed, is carefully explained in the text. Most students find the Boltzmann Law to be mathematically the most demanding part of the book, since it involves exponentials and natural logarithms. As far as I can judge, students have no more difficulty with this section than they have with pH.

It has been my aim to make the argument in this book plausible rather than rigorous — to provide an initial intuitive insight rather than a formal logical structure. The place for deriving thermodynamics from a minimum number of hypotheses or deducing an unexceptionable scale of temperature is in more advanced treatments and not in a freshman text.

The subject matter contained in this book is designed to be covered toward the end of the freshman year in the second semester or third quarter. The author assumes that the reader is acquainted with most of freshman chemistry, including equilibrium constants and the elements of heat flow. Students without this background can find a concise summary of these subjects in Appendices 1 and 2.

As seems inevitable in a supplementary text, there is more material here than can conveniently be covered in the time available. I have found it possible to cover the contents of the first nine chapters in about twelve lectures only by judiciously omitting some subject matter, particularly in the first three chapters.

Among the many friends and colleagues with whom I have had discussions in writing this book there are four people to whom I am particularly indebted for advice and assistance. I would like to thank Dr. John Bradley of the University of the Witwatersrand for his quiet enthusiasm for my initial ideas. Without his encouragement, I might well not have begun to write this book. I must thank Dr. Emil Slowinski of Macalester College for several major suggestions which have resulted in an overall improvement in the structure of the book, and also Dr. William Masterton of the University of Connecticut both for some acute criticisms and for extensive improvements to the style. Additionally, I thank Dr. Dorothy Feigl of St. Mary's College not only for some very able proofreading but also for her constant and friendly encouragement. Finally I wish to acknowledge my indebtedness to the University of the Witwatersrand for granting me six months' paid leave during which many of the central ideas in this book took shape.

WILLIAM G. DAVIES

CONTENTS

Chapter 1

EQUILIBRIUM AND PROBABILITY

INTRODUCTION

Most beginning students in chemistry quickly develop an intuition that there is an "uphill" and a "downhill" character to chemical reactions. When, for example, sodium reacts with chlorine to form sodium chloride, according to the equation

$$2 \, Na(s) + Cl_2(g) \rightarrow 2 \, NaCl(s) \qquad \textbf{(1.1)}$$

the reaction proceeds of its own accord without any outside assistance. In this respect, at any rate, it behaves like a stone rolling down a hillside. There is a "downhill" character to reaction (1.1).

By contrast, there is an "uphill" character to the reverse of reaction (1.1). Sodium chloride crystals do not decompose of their own accord into sodium metal and chlorine gas. It is possible to decompose sodium chloride into its elements, of course, but only by involving outside assistance. We could, for instance, melt the crystals and electrolyze them. It is also possible to get a stone from the bottom of the hill to the top, but only by involving outside assistance; e.g. by rolling it up the hillside by hand. Stones do not roll uphill on their own, nor do sodium chloride crystals decompose on their own.

This "uphill" and "downhill" character to chemical reactions is not confined to the reaction of sodium and chlorine, but applies to *all* reactions. When hydrogen reacts with oxygen to form water, for instance, such a reaction obviously has a downhill character to it, since its reverse, the decomposition of water into its elements, can only be brought about by some outside agency. Similarly, we would attribute a downhill character to such familiar chemical reactions as the oxidation of magnesium, the reaction of an acid with a base, the reaction of hydrogen with chlorine, and so on.

The question now arises, whether this intuition about an uphill-downhill character to chemical reactions has some basis in reality. All of us know that to have a hunch is not enough. Not all hunches pay off. Yet this particular one is worth investigating, since it leads us immediately to ask a very important question. If there is an uphill-downhill aspect to chemical reactions, there must be a *reason* for such behavior. Is there then a basic property common to all chemical reactions, explaining why those reactions that do occur can occur, and why those that do not occur cannot occur? If there is such a property, then it is one of the first things we should learn about when we study chemistry.

The study of these and similar questions falls within the boundaries of two rather different disciplines, one of which is called *statistical mechanics* and the other *thermodynamics*. Statistical mechanics is concerned with interpreting the behavior of matter from a *microscopic* viewpoint; i.e., in terms of the average behavior of a very large number of molecules. Thermodynamics, on the other hand, regards matter from a *macroscopic* viewpoint. It deals with the bulk properties of matter rather than the properties of individual molecules.

In this book we will purposely not confine ourselves to using a purely statistical or a purely thermodynamic argument, but use whichever suits our purposes best. At first our approach will be largely in terms of molecular behavior, since beginning students find such an approach to chemical equilibrium more easy to grasp. Later, when we have appreciated the essentials of what is at stake in chemical equilibrium, we will be in a position to switch to a more convenient but less molecular approach involving the bulk properties of matter.

Although thermodynamics and statistical mechanics have many useful applications in physics and engineering, in this book we will confine ourselves to chemical applications. In particular, we will center our discussion around the question just raised, namely, when chemical reactions can occur and when they cannot. It is chiefly because thermodynamics and statistical mechanics can answer this question that chemists are interested in these subjects.

EQUILIBRIUM

In talking about the uphill and downhill character of chemical reactions, all the examples we considered were of reactions which went virtually to completion. They were, if you like, reactions which went downhill all the way. All the reactants were consumed and converted into products. No detectable amounts of reactants were left at the end unless they were initially present in excess. The "bottom of the hill," so to speak, corresponded to the pure products.

Such a situation does not always occur. Sometimes the reaction does not go to completion and only a portion of the reactants is transformed into products. Eventually we are left with a situation in which appreciable amounts of both reactants and products are present. Here the "bottom of the hill" analogy does not correspond to pure reactants or pure products but to some intermediate state.

An example of such a situation is the reaction

$$H_2(g) + D_2(g) \rightarrow 2\ HD(g). \tag{1.2}$$

If a mole of hydrogen gas and a mole of deuterium gas are mixed in a container in the presence of a suitable catalyst, such as powdered platinum metal, the two gases react to form hydrogen deuteride, HD. With the passage of time the concentration of hydrogen deuteride increases while that of hydrogen and deuterium decreases. This process continues until the amount of hydrogen deuteride formed is one mole. This is exactly half of the quantity which would be formed if reaction (1.2) had gone to completion. Thus the final situation is not two moles of pure HD but an intimate mixture of one mole of HD with half a mole of H_2 and half a mole of D_2.

Once this final situation has been achieved, no further concentration changes are observed to occur with time. Such a final situation is said to be an *equilibrium state.* The one mole of HD, the half-mole of H_2 and the half-mole of D_2 are together said to constitute the *equilibrium mixture,* and to be in *equilibrium with each other.*

This same final equilibrium mixture can be obtained in another way. If two moles of HD are taken, and the same catalyst used as above, the HD dissociates, forming some H_2 and D_2, according to the equation

$$2\ HD(g) \rightarrow H_2(g) + D_2(g). \tag{1.3}$$

The dissociation does not go to completion. Once one of the two moles of HD has dissociated, no further changes in concentration take place. The final equilibrium mixture is the same as that obtained when H_2 and D_2 were mixed, namely one mole of HD and half a mole each of H_2 and D_2. Thus it does not matter whether we start with two moles of HD or one each of H_2 and D_2, the final equilibrium mixture is the same.

We can also look at this reaction system in terms of the "uphill" and "downhill" property postulated above. If such a property exists, then the situation must be something like that depicted in Figure 1-1. Whether we start with 1 mole each of H_2 and D_2 and no HD as at the left of the figure, or with 2 moles of HD and none of H_2 or D_2 as at the right of the figure, we end up with 1 mole HD and half a mole each of H_2 and D_2 at the center of the figure. The stone of our analogy can roll down to the bottom of the valley from either side. The bottom of the curve corresponds to the equilibrium position.

The hydrogen-deuterium equilibrium state exhibits two features which are essential to *any* equilibrium state. Firstly, an equilibrium states does not change with time; and secondly, an equilibrium state can be attained from more than one direction. If either of these two features is lacking, then the system we are studying is not in an equilibrium state.

It is sometimes very difficult to recognize an equilibrium state. In particular,

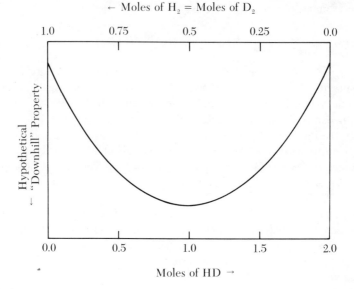

Figure 1-1. A possible way of regarding the hydrogen-deuterium equilibrium.

many chemical systems give the appearance of not changing with time, but are not really in an equilibrium state. Such systems are in fact changing so slowly that we cannot readily observe the change. An example of such an apparent state of equilibrium is a mixture of two moles of hydrogen gas and one mole of oxygen gas. Such a mixture can be left at room temperature in the absence of a catalyst for a whole lifetime without any detectable changes taking place. If we had no other evidence at hand, we could easily conclude that the hydrogen-oxygen mixture was in an equilibrium state. That this is not the case becomes clear once we try to attain this apparent equilibrium state by some other route than by simply mixing the two constituent gases. We could, for instance, try to persuade water to decompose of its own accord into its elements, perhaps by using an appropriate catalyst, or by striking a spark in it. Such a search would meet with no success, and would suggest that it is water rather than a hydrogen-oxygen mixture which corresponds to the equilibrium state. A further search for catalysts would reveal that there are several which will allow hydrogen and oxygen to react with explosive violence to form water. Despite the appearance of not changing with time, a hydrogen-oxygen mixture is thus not really in an equilibrium state.

    Although it does nothing to solve the problem of the uphill and downhill character of chemical reactions, the introduction of the concepts of equilibrium and equilibrium state at least provides the right vocabulary with which to discuss the problem. What we have called an equilibrium state obviously corresponds to a "bottom of the hill" situation. By definition, an equilibrium state is one in which no change occurs with time or can occur with time. Only if a chemical system is in an "out-of-equilibrium" state will changes tend to occur. Such changes as do occur, moreover, will bring the system closer to an equilibrium state. The reverse of such a process does not occur in nature without outside help. Once a system is in an equilibrium state, it does not move of its own accord to an out-of-equilibrium state. Thus the "downhill" character of a reaction corresponds to an

out-of-equilibrium → equilibrium

process, while the "uphill" character of a reaction corresponds to the reverse, to an

equilibrium → out-of-equilibrium

process.

Seen in these terms, the problem of predicting the direction of a chemical reaction becomes one of finding a criterion for equilibrium. If we can only discover the basic reason for the existence of an equilibrium state, it should then be possible to tell whether a given chemical system is in an equilibrium state or not, and if it is not, in which direction equilibrium lies. Any reaction which takes place will proceed toward the equilibrium state.

Is there such a basic criterion for equilibrium? The answer is yes; there is a very simple criterion. It reads as follows: **An equilibrium state is the most probable state** which can be attained by the reaction system. Conversely, an out-of-equilibrium state corresponds to a *less probable* (in most cases a *highly improbable*) situation. Such a proposition is, of course, a hypothesis; that is, it is a basic assumption which cannot be deduced from more basic assumptions. It can only be vindicated by experiment. And vindicated it has been! Deductions from this hypothesis have *never yet* failed to be valid in all the accumulated experience of chemistry.

Before we can appreciate the significance of this hypothesis, we must first familiarize ourselves with the idea of probability, and how we can measure it as a number. Not until we have done this can we begin to gain a real insight into the problem of chemical equilibrium.

PROBABILITY

The probability of an event occurring is usually measured by a number between 0 and 1, to which we give the symbol p. The more likely an event is to occur the larger will be the value of p. An event which is *certain* to occur has the highest possible value of p, namely 1. By contrast an event which will *never* occur is given the lowest possible value of p, namely 0.

The probabilities of interest to us in our study of equilibrium are those which derive from *equally likely events*. Such probabilities are easy to deal with numerically. An example of such a probability is that of drawing a particular card from a well shuffled deck of the usual 52 cards. The probability of drawing a particular card, say the king of diamonds, is given by

$$p = \frac{1}{52}.$$

We assume that it is *equally likely* that we can draw any one of the 52 cards in the deck. Of these 52 equally likely possibilities, only one possibility is of the kind in which we are interested.

Suppose, by contrast, that we wish to calculate the probability of drawing a spade from the deck. Of the 52 equally likely possibilities, there are 13 of the kind in which we are interested, namely spades. The probability of drawing a spade is then given by

$$p = \frac{13}{52} = \frac{1}{4}.$$

In general, if an event in which we are interested can occur in N equally likely ways, of which m are of a particular kind, then the probability of an "m type" event is given by

$$p = \frac{m}{N}. \tag{1.4}$$

Having defined the probability of an event in terms of a number, let us now ask what significance such a number has. What significance does it have, for instance, to say that the probability of drawing a spade from a deck of cards is 1/4? The number 1/4 has the following significance: If we draw a card from a deck a very large number of times, we will find that a spade will be obtained approximately one fourth of the times. The longer we go on with such a process, the closer the result will be to exactly one fourth.

INDEPENDENT EVENTS

Two events are said to be *independent* if the occurrence of the first event has no effect on the probability of the occurrence of the second. Consider, for instance, successive flips of a coin. When a normal coin is flipped, we expect heads or tails to be equally likely. The probability of the coin falling heads is thus

$$p = \frac{1}{2}.$$

If we now pick up the same coin and flip it again, the same probability situation arises. On a second flip, the probability of the coin falling heads is still 1/2. The same will be true of any subsequent flip of the coin. Each flip of the coin is said to be *independent* of the others.

Next let us consider a case where the probabilities are *not* independent. If we draw cards from a deck without returning them to the deck each time, then the successive draws are not independent of each other. The probability of obtaining a spade on the first draw is 13/52, as we have just seen. The probability of obtaining a spade on the second draw, though, is *not* 13/52. It is 12/51 if the first card was a spade, and 13/51 if the first card was not a spade. The occurrence of the first draw thus affects the probability of the second, and the two events are not independent.

We must next consider the probability of several independent events occurring together. We take a simple example: flipping the same coin twice. What is the probability of heads occurring on the first flip, and heads occurring on the second flip *as well?* Indicating H for heads and T for tails, and 1 and 2 for the flips, we have a total of four equally likely possibilities:

Flip 1	*Flip 2*
H	H
H	T
T	H
T	T

Of these four possibilities, only one, the first, is of the required type. Thus the probability of flipping heads twice in succession is given by

$$p = \frac{1}{4}.$$

The result just obtained is a particular example of a general theorem in probability theory, the *law of multiplication of independent probabilities*. If p_1 is the probability of one event, p_2 that of a second event, p_3 that of a third, and so on, and all the events are independent of each other, then the probability of all the events occurring in a specified sequence is the product of their independent probabilities; i.e., the probability is given by:

$$p = p_1 \times p_2 \times p_3 \times \ldots \ldots \quad \text{(1.5)}$$

Thus, in the example just discussed, the probability of achieving heads on the first flip is 1/2, and the probability of achieving heads on the second is likewise 1/2. The probability of heads being achieved *both* on the first *and* on the second flip is thus $1/2 \times 1/2 = 1/4$, as discovered without resorting to the theorem.

A further example will make this theorem clearer. A student takes a multiple choice test consisting of thirty questions. He is given five choices: A, B, C, D, and E for each question. What is the probability that he can answer all thirty questions correctly just by blind guessing?

Since his guess for any one question will have no bearing on his guess for any other question, we can regard the probability of each guess as being independent of the others. Also, since the guessing is supposed to be "blind" rather than "educated," we can assume that the probability of the student obtaining the correct answer for a given question is $1/5$. The problem of guessing all thirty answers correctly is thus $1/5$ multiplied by itself thirty times, or

$$p = \left(\frac{1}{5}\right)^{30}.$$

To evaluate this, we take the logarithms of both sides of the equation:

$$\log p = 30 \log \frac{1}{5} = -30 \log 5$$

$$= -30 \times 0.70 = -21$$

$$p = 1 \times 10^{-21}$$

This is a very small probability indeed. We can expect a student to achieve a perfect set of answers purely by chance only about once in every 10^{21} tests. We would have to test the whole human race once a week for some ten billion years to achieve this number of tests! An instructor would surely be justified in assuming that a perfect set of answers was not obtained by chance.

Although 10^{-21} is a very small probability, it is very large compared to some of the probabilities we shall shortly encounter!

THE H_2-D_2-HD EQUILIBRIUM

Now that we have learned a little about probability, we can apply our knowledge to a situation of chemical interest. We will investigate a system discussed previously, namely the equilibrium between H_2 gas and D_2 gas on the one hand, and HD gas on the other. We indicate such an equilibrium as follows:

$$H_2(g) + D_2(g) \rightleftharpoons 2\,HD(g) \tag{1.6}$$

The double arrows indicate an equilibrium situation in which appreciable quantities of both reactants and products are present.

As before, we assume that we start with a mole each of H_2 and D_2 gases and allow them to react in the presence of an appropriate catalyst. At first, the concentrations of H_2 and D_2 will decrease, while the concentration of HD will increase. Finally, an equilibrium situation will be attained in which all concentrations remain constant.

When viewed in terms of concentration, such an equilibrium situation appears static, but on a molecular level everything is in a state of flux. The molecules are not only continually interchanging energy, but also continually interchanging atoms. A particular H atom could, for instance, be joined at one instant to a D atom. A little later on, the same H atom might be joined to another H atom. Later still, it could be joined to a different D atom. No H or D atom retains a given partner for very long.

In such a constant reshuffle of atoms among molecules at equilibrium, notice that both *forward and reverse* reactions are occurring simultaneously. Not only are H_2 and D_2 molecules reacting to produce HD molecules according to the equation

$$H_2(g) + D_2(g) \rightarrow 2\,HD(g) \qquad \textbf{(1.2)}$$

but HD molecules are reacting to produce H_2 and D_2 molecules at the same time:

$$2\,HD(g) \rightarrow H_2(g) + D_2(g). \qquad \textbf{(1.3)}$$

Furthermore, both processes are occurring *at the same rate.*

As fast as HD molecules are being produced by process (1.2), they are being destroyed by process (1.3). This must be the case if the concentration of HD is to remain constant with time. Such a situation, in which a continuous balance between opposing processes produces a static result, is usually referred to as a *dynamic equilibrium.*

Before we can apply simple probability theory to this equilibrium, we must make one crucial assumption. We must assume that although hydrogen and deuterium atoms have different masses, they behave identically in this constant reshuffling of partners. In other words, there is no tendency for H to associate with H rather than with D, or for D to associate with D rather than with H.

Since our mixture contains equal numbers of H atoms and D atoms, it follows that if we select a molecule at random from the mixture and label one of its constituent atoms 1, this atom is just as likely to be an H atom as it is to be a D atom. It is also true that the *other* atom in the molecule, atom 2, is just as likely to be an H atom as it is to be a D atom. Our randomly selected molecule is thus equally likely to be one of four possibilities:

Atom 1	Atom 2
H	H
H	D
D	D
D	H

The probability of each possible combination is 1/4. The situation is, in terms of probability, identical to that of two successive flips of a coin.

Of the four equally likely possibilities, two correspond to the mixed species HD. Thus the probability of a randomly selected molecule being an HD molecule is 1/2, while that of it being an H_2 molecule is 1/4. The probability of it being a D_2 molecule is likewise 1/4. In the equilibrium mixture of constantly interchanging partners we therefore expect that at any given time, half the molecules will be HD molecules, a quarter will be H_2 molecules, and a quarter will be D_2 molecules. Since we started with a mole each of H_2 and D_2 gases, on the basis of probability alone we expect the equilibrium mixture to consist of a mole of HD and half a mole each of H_2 and D_2.

THE EQUILIBRIUM CONSTANT FOR THE H₂–D₂–HD EQUILIBRIUM

The equilibrium constant* for the reaction

$$H_2(g) + D_2(g) \rightleftharpoons 2\,HD(g) \tag{1.6}$$

is given by the expression

$$K_c = \frac{[HD]^2}{[H_2]\,[D_2]}. \tag{1.7}$$

It is very easy to calculate the value of K_c for this reaction from the composition of the equilibrium mixture we have just deduced. If the total volume of the equilibrium mixture is V liters, then the concentrations of the three species concerned are given by:

$$[HD] = \left(\frac{1}{V}\right) \text{mole liter}^{-1}$$

and

$$[H_2] = [D_2] = \frac{\frac{1}{2}}{V} \text{ mole liter}^{-1}.$$

---

*Readers unfamiliar with the equilibrium constant will find a brief account in Appendix 1.

Inserting these values into Equation (1.7), we have

$$K_c = \frac{\left(\frac{1}{V}\right)^2}{\left(\frac{1}{2V}\right)\left(\frac{1}{2V}\right)} = 4$$

If our theory is correct, we should find that K_c for this reaction has the value of 4 independently of temperature. Is such a prediction borne out by experiment? The H_2-D_2-HD equilibrium is not an easy one to study experimentally, but it has been investigated. The experimental results agree surprisingly well with the predicted value. K_c is found to be a little smaller than 4, and to rise with temperature. At 25°C, for instance, K_c is found to have the value 3.3, and at 500°C it has the value 3.8. The discrepancy between theory and experiment can be attributed to the approximation we have made in assuming that H and D atoms behave identically. In fact, there is a slight difference in behavior. Nevertheless, the agreement between theory and experiment is sufficiently close to suggest that the broad outline of our theory is correct even if the details are slightly wrong. We have established that there is a connection between *equilibrium* on the one hand and *probability* on the other.

Up to the present, we have only considered the case in which there are equal numbers of H and D atoms in the equilibrium mixture. It is instructive to extend the arguments to cases in which this is not so. Suppose, for instance, that we have an equilibrium mixture prepared by adding x moles of H_2 gas to y moles of D_2 gas in a container of V liters in the presence of a catalyst.

In such an equilibrium mixture, there are 2x moles of H atoms, some attached to D atoms and some to other H atoms. Similarly, there are 2y moles of D atoms. The total number of moles of atoms of both kinds is (2x + 2y) moles. Let us now focus our attention on a single molecule selected at random from the equilibrium mixture, and label its atoms 1 and 2.

The probability of atom 1 being an H atom is

$$p_H = \frac{\text{number of moles of H}}{\text{number of moles of H} + \text{number of moles of D}} = \frac{2x}{2x + 2y} = \frac{x}{x + y}.$$

The probability of atom 1 being a D atom is, similarly,

$$p_D = \frac{2y}{2x + 2y} = \frac{y}{x + y}.$$

(Notice that the two probabilities add up to unity, a convenient cross check. Atom 1 *must* be H or D.) The same arguments used in the case of atom 1 also apply to atom 2. Thus, the probability of atom 2 being H is also x/(x+y), and the probability of it being D is $y/(x+y)$.

By multiplying probabilities, we can now find the probability of each of the four possible combinations of H and D for the two atoms in the molecule. We find:

Possible Combination		Probability
Atom 1	Atom 2	
H	H	$\dfrac{x}{x+y} \cdot \dfrac{x}{x+y} = \dfrac{x^2}{(x+y)^2}$
H	D	$\dfrac{x}{x+y} \cdot \dfrac{y}{x+y} = \dfrac{xy}{(x+y)^2}$
D	H	$\dfrac{y}{x+y} \cdot \dfrac{x}{x+y} = \dfrac{xy}{(x+y)^2}$
D	D	$\dfrac{y}{x+y} \cdot \dfrac{y}{x+y} = \dfrac{y^2}{(x+y)^2}$

We can check these results in two ways. Firstly, the sum of all four probabilities is

$$\frac{x^2 + 2xy + y^2}{(x+y)^2} = 1;$$

i.e., it is certain that the molecule must be one of the four possibilities shown. A second cross-check is obtained by making $x = y$. This gives exactly the same result as obtained above for equal quantities of H and D atoms, as Table 1-1 shows.

Table 1-1

Species	Probability (general case)	Probability $(x = y)$
H-H	$\dfrac{x^2}{(x+y)^2}$	$\dfrac{x^2}{4x^2} = \dfrac{1}{4}$
H-D	$\dfrac{xy}{(x+y)^2}$	$\dfrac{x^2}{4x^2} = \dfrac{1}{4}$
D-H	$\dfrac{xy}{(x+y)^2}$	$\dfrac{x^2}{4x^2} = \dfrac{1}{4}$
D-D	$\dfrac{y^2}{(x+y)^2}$	$\dfrac{x^2}{4x^2} = \dfrac{1}{4}$

The probabilities we have deduced tell us that, in the equilibrium mixture of x + y moles of molecules, we can expect to find that

$$\frac{x^2}{x + y} \text{ moles will be } H_2$$

$$\frac{y^2}{x + y} \text{ moles will be } D_2$$

$$\frac{2xy}{x + y} \text{ moles will be HD}$$

If V is the total volume of the equilibrium mixture, then the equilibrium concentrations are:

$$[H_2] = \frac{x^2}{x + y} \cdot \frac{1}{V} \text{ mole liter}^{-1}$$

$$[D_2] = \frac{y^2}{x + y} \cdot \frac{1}{V} \text{ mole liter}^{-1}$$

and

$$[HD] = \frac{2xy}{x + y} \cdot \frac{1}{V} \text{ mole liter}^{-1}$$

Substituting these concentrations into the expression for the equilibrium constant, equation (1.7), we find that

$$K_c = \frac{\dfrac{(2xy)^2}{(x + y)^2 \cdot V^2}}{\dfrac{x^2}{(x + y)V} \cdot \dfrac{y^2}{(x + y)V}} = \frac{4\,x^2 y^2}{x^2 y^2} = 4.$$

Our probability argument thus predicts that whatever our initial mixture, i.e., no matter what the values of x, y, and V are, we will *always* find that at equilibrium the expression

$$\frac{[HD]^2}{[H_2]\,[D_2]}$$

has a constant value, namely 4.

The invariance of the equilibrium constant K_c is thus a natural outcome of looking at this equilibrium in terms of probability.

FLUCTUATIONS

If someone were to flip a coin one thousand times in succession, we would not be altogether surprised if the result were 498 heads and 502 tails, rather than 500 of each. Such a result would reflect neither on the balance of the coin nor on the laws of probability. The truth is that although the 500-500 situation is the most probable outcome, the 498-502 situation is very nearly as probable.

We have a similar situation with the H_2-D_2-HD equilibrium just discussed. If equal numbers of H_2 and D_2 molecules are mixed and allowed to establish equilibrium, we do not expect *exactly* half of the molecules to be HD molecules *forever* afterwards. Rather, we expect the number of HD molecules to fluctuate slightly, being somewhat larger than an exact half at times and somewhat smaller at other times. How big will these fluctuations be? The question is well worth investigating, since in the process of answering it we come to realize that while small fluctuations are always occurring, really large fluctuations, big enough to be observable, are *inconceivably improbable.* This result is important to our understanding of equilibrium. It enables us to appreciate not only that an equilibrium state is the most probable state but that is is *overwhelmingly more probable* than any other eventuality.

We will start our investigation of fluctuations by considering an example in which only a few molecules are involved. The calculation in such a case is fairly straightforward, and it gives us an intuitive feeling for what happens when a larger number of molecules is involved, where the calculations become too complex for us to attempt.

The system we will consider is one in which there is a total of four molecules, involving four H atoms and four D atoms. If these four molecules are constantly interchanging partners, we have three possible situations which are chemically distinct. These are:

$$\text{(a)} \quad 2H_2 + 2D_2$$

$$\text{(b)} \quad 2HD + H_2 + D_2$$

$$\text{(c)} \quad 4HD$$

No other possibility is consistent with four molecules, four H atoms, and four D atoms.

Intuitively, we realize that (a), (b), and (c) are not all equally probable, and that (b) is the most probable of the three; but it is not easy to see *why* this should be so. In order to see this, we must first devise a scheme for labelling molecules and atoms.

Suppose that it is possible to photograph the four gas molecules of our system at a particular instant of time. A possible result of such a photograph is shown in Figure 1-2. In this figure, each molecule is numbered and its left-hand

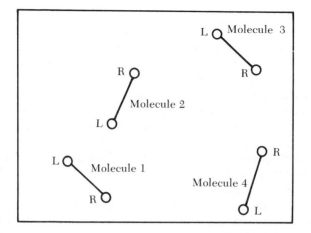

Figure 1-2. An instantaneous photograph of four diatomic gas molecules.

and right-hand sides are indicated. There is a total of eight positions in the photograph. Of these eight positions, four must be occupied by H atoms and four by D atoms.

Table 1-2

Atom	Mol 1 L–R	Mol 2 L–R	Mol 3 L–R	Mol 4 L–R
1	H–H	H–H	D–D	D–D
2	H–H	H–D	H–D	D–D
3	H–H	H–D	D–D	D–D
4	H–H	H–D	D–D	H–D
5	H–H	H–D	D–D	D–H
6	H–D	H–H	H–D	D–D
7	H–D	H–D	H–D	D–D
8	H–D	H–D	H–D	H–D
9	H–D	H–D	H–D	D–H
10	H–D	H–D	D–H	H–D
11	H–H	D–H	H–D	H–D
12	H–D	D–D	H–D	D–H
13	H–D	D–D	D–H	H–D
14	H–D	D–D	D–H	D–H
15	H–D	D–D	D–D	H–H
16	H–D	H–H	H–D	D–D
17	H–D	H–H	H–D	H–D
18	H–D	H–H	H–D	D–H
19	H–D	H–H	H–D	D–D
20	H–D	H–D	H–H	H–D
21	H–D	H–D	H–D	H–D
22	H–D	H–D	H–D	D–H
23	H–D	H–D	D–D	H–H
24	H–D	H–D	D–D	H–D
25	H–D	H–D	D–D	H–H
26	H–D	D–H	H–H	D–D
27	H–D	D–D	H–H	D–D
28	H–D	D–D	H–D	D–H
29	H–D	D–D	H–D	D–H
30	H–D	D–D	H–D	H–D
31	H–D	D–D	H–D	D–H
32	H–D	D–D	D–H	H–D
33	H–D	D–D	D–H	H–D
34	H–D	D–D	D–H	H–H
35	H–D	D–D	D–D	H–H
36	D–D	D–D	D–H	H–H
37	D–D	D–D	H–D	H–H
38	D–D	D–D	H–H	D–H
39	D–D	D–D	H–H	H–D
40	D–D	D–D	H–H	H–D
41	D–D	H–D	D–D	H–H
42	D–D	H–D	H–D	D–H
43	D–D	H–D	H–H	D–H
44	D–D	H–D	H–H	H–D
45	D–D	H–D	H–D	D–H
46	D–D	H–D	H–D	H–D
47	D–D	H–D	H–D	H–H
48	D–D	H–H	H–D	D–H
49	D–D	H–D	H–H	D–D
50	D–D	H–H	H–H	D–D
51	D–H	D–D	D–H	H–H
52	D–H	D–D	H–D	H–H
53	D–H	D–D	H–D	H–H
54	D–H	D–D	H–H	H–D
55	D–H	D–H	D–D	H–H
56	D–H	D–H	D–H	D–H
57	D–H	D–H	D–H	H–D
58	D–H	D–H	H–H	D–D
59	D–H	D–H	H–H	D–H
60	D–H	D–H	H–H	H–D
61	D–H	H–D	D–D	H–H
62	D–H	H–D	D–D	H–H
63	D–H	H–D	D–H	H–D
64	D–H	H–D	H–H	D–D
65	D–H	H–D	H–D	H–D
66	D–H	H–D	H–D	H–D
67	D–H	H–H	D–D	D–H
68	D–H	H–H	D–D	H–D
69	D–H	H–H	H–D	H–D
70	D–H	H–H	H–D	D–D

We now inquire as to how many ways there are in which the four H atoms and four D atoms can possibly be arranged in the photograph. The answer is larger than one might suppose. There are a total of seventy different ways! All seventy of these are listed in Table 1-2. To make sure that we understand the significance of this table, consider the arrangement numbered 12. This corresponds to a photograph in which molecule 1 is H_2, molecule 2 is D_2 and molecules 3 and 4 are both HD molecules. In molecule 3, the *left*-hand atom is H, while in molecule 4 the *right*-hand atom is H. Such a photograph is indicated in Figure 1-3.

It is not difficult to argue that each of the seventy possibilities listed in Table 1-2 is equally likely to appear in our hypothetical photograph. This result follows from the assumption that H and D atoms behave identically. Let us consider arrangement 12 shown in Figure 1-3 again, as an illustration of this point. Suppose that the H and D atoms in molecule 3 are interchanged, so that the molecule is D-H rather than H-D. This new arrangement (which is actually number 14 in Table 1-2) will be just as probable as the original one shown in the figure. In a similar way, we can argue that *any* two arrangements which can be obtained from each other by interchanging an H and a D atom are equally probable. Since it is possible to derive any of the 70 arrangements in the table from any other arrangement by a sufficient number of H-D interchanges, all 70 must be equally probable.

If all 70 arrangements are equally probable we can now determine the probabilities of the three distinct chemical situations, (a), (b), and (c), considered above. All we need do is find out how many of the arrangements in Table 1-2 correspond to each situation. We discover that numbers 1, 10, 15, 36, 45, and 50 correspond to two H_2 and two D_2 molecules, or situation (a). Further

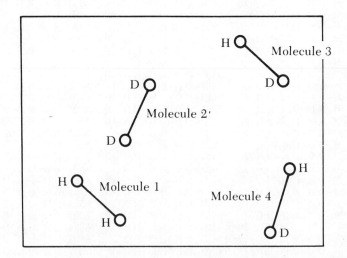

Figure 1-3. A "photograph" of arrangement 12 of Table 1-2, showing the positions of the individual atoms.

inspection reveals that 16 arrangements correspond to four HD molecules, or situation (c); while the remaining 48 arrangements correspond to situation (b).
Tabulating these results, we find:

Composition	Number of arrangements	Probability
(a) $2 H_2 + 2 D_2$	6	$\frac{6}{70} = 0.0857$
(b) $2HD + H_2 + D_2$	48	$\frac{48}{70} = 0.6857$
(c) 4 HD	16	$\frac{16}{70} = 0.2286$

The hypothetical photograph we have been discussing could have been taken at any instant in time without affecting our argument. Thus the probabilities just listed tell us the likelihood of each of the three compositions, (a), (b), and (c), occurring not just at *one* specific time but at *any* instant of time. In other words, they tell us that the system will have the composition $2H_2 + 2D_2$ for 8.57 per cent of the time, the composition $2HD + H_2 + D_2$ for 68.57 per cent of the time, and the composition 4 HD for 22.86 per cent of the time.

The four-molecule system we have been discussing thus behaves much as we would expect. The composition is not constant with time, but fluctuates around the most probable composition. This most probable composition is, furthermore, the same as the equilibrium composition obtained above for a large number of molecules, namely 50 per cent HD.

If we now try to expand the approach we have used for a four-molecule system to deal with systems involving larger numbers of H_2, D_2, and HD molecules, we immediately run into a difficulty. We cannot easily make a list like Table 1-2 which includes all the possible ways in which H and D atoms can be arranged among the molecules. Even for four molecules, such a list is a bit cumbersome. Instead we must develop formulas which apply to any number of molecules in order to calculate probabilities. The development of such formulas is a little beyond the scope of this book, and we shall not attempt it here. Instead we shall quote the results of such calculations.

Table 1-3 and Figure 1-4 show how the probabilities behave as the total number of molecules in a system rises. Results are given for 16, 100, and 1,000 molecules. All the cases quoted refer to systems in which there are equal numbers of H and D atoms. The most probable situation is thus always 50 per cent HD, 25 per cent H_2, and 25 per cent D_2, irrespective of the number of molecules. In both the table and the figure, the probabilities quoted are relative rather than actual probabilities: the probability of the 50 per cent mixture is taken as unity, and the other probabilities are quoted relative to this. For example, in the case n = 100, the probability of 38 per cent HD (i.e., 38 HD, 31 H_2, and 31 D_2 molecules) is quoted as 0.051. This means that in the constant

Table 1-3
Relative probability, rp, of finding x per cent HD in a mixture of n
molecules made up from equal numbers of H and D atoms.

n = 16		n = 100		n = 1000	
x	rp	x	rp	x	rp
0	0.000	0	0.000	0	0.000
12.5	0.007	26	0.000	40	0.000
25.0	0.117	38	0.051	47	0.161
37.5	0.560	46	0.635	48	0.441
50.0	1.000	48	0.905	49	0.881
62.5	0.711	50	1.000	50	1.000
87.5	0.017	52	0.943	51	0.827
100.0	0.000	62	0.064	52	0.459
		74	0.000	53	0.171
		100	0.000	55	0.007
				100	0.000

reshuffle of partners, 38 per cent HD will turn up only 0.051 times as often as
will 50 per cent HD.

Both the table and the figure demonstrate a very obvious trend as the
number of molecules increases. The gently rounded peak corresponding to n =
16 in Figure 1-4 becomes narrower as n increases. By the time n reaches the

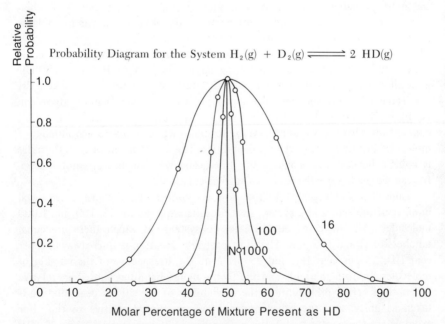

Probability Diagram for the System $H_2(g) + D_2(g) \rightleftharpoons 2\ HD(g)$

Figure 1-4. As the number of molecules in the system rises from 16 to Avogadro's number
(N), the peak of the probability curve becomes very narrow, centering on 50 per cent HD.

Avogadro* number N, the peak has become a needle very much finer than the actual line printed in the figure. In other words, for small numbers of molecules the probability of finding the system with a percentage of HD somewhat different from 50 per cent is quite appreciable, but as larger and larger numbers become involved the probability of such "way out" distributions diminishes drastically. With numbers as large as the Avogadro number, any percentage of HD markedly different from 50 per cent has so minute a probability that we can safely discount the possibility of it ever being observed.

For example, it is possible to calculate the probability of finding 50.00001 per cent HD in a mole of equilibrium mixture relative to that of finding *exactly* 50 per cent HD. The answer is such a small number that it is difficult to write it down! It is the number:

$$\frac{1}{10^{10,000,000,000}}$$

i.e., it is one divided by ten to the power ten billion! An event with such an inconceivably small probability is unlikely to occur not only in our own lifetime, but in the whole lifetime of the universe. For all practical purposes we can regard it as being impossible.

A composition of 50.00001 per cent HD may seem to be very close to a composition of exactly 50 per cent HD when we think in terms of percentages, but when we think in terms of molecules the difference is quite a large number. While exactly 50 per cent HD corresponds to exactly $N/2$ molecules of HD, a composition of 50.00001 per cent HD corresponds to $N/2 + (6 \times 10^{16})$ molecules. That this particular composition is so improbable thus means that a fluctuation of 6×10^{16} molecules from the ideal is also extremely improbable. Presumably, though, *some* fluctuations occur. How large or how small do such fluctuations have to be in order to occur with any frequency?

The answer is that a fluctuation as large as 10^{11} or 10^{12} molecules is almost as probable as a fluctuation of only one or two molecules. Fluctuations much larger than 10^{12}, though, are extremely unlikely. Such a result means that if we had some device for counting the number of HD molecules in the mixture at any instant of time and counted them, say, every microsecond, we would usually find an excess or a deficiency of some 10^{11} or 10^{12} HD molecules. Occasionally, about once a day, we would find a really small fluctuation of one or two, or even zero, molecules. By contrast, we could wait till the end of time before observing a fluctuation as large as 10^{16} molecules.

The fact that fluctuations occur does not, of course, destroy the meaning of what we have already said about an equilibrium state. Even though the ideal composition of an *exact* half mole of HD occurs very rarely, it still makes sense

---

*The Avogadro number, named for Amadeo Avogadro, is the number of molecules in a mole. It is 6.02×10^{23}.

to talk about an equilibrium composition of the mixture. A fluctuation of 10^{12} molecules may be a large number of *molecules,* but it is a very small *fraction* of the total number of HD molecules present in the equilibrium mixture; in fact, it is only about 10^{-11} of the total. In order to detect such a fluctuation we would need some analytical technique which could determine the concentration of HD to 11 significant figures. No such analytical technique exists. In other words, the fluctuations we have been talking about are so small as to make no difference in measured concentrations.

What we have said about the fluctuations in composition of a system in an equilibrium state can also be applied to other properties of the system, in particular to its temperature and pressure. If we keep our equilibrium mixture at $0°C$ and 1 atmosphere in a vessel of appropriate volume, the theoretical number of molecules hitting one square centimeter of wall per second is 3.25×10^{24}. If we were able to count the exact number striking a given square centimeter per second for a large number of seconds, we would count a different number each second. Nevertheless, such fluctuations in numbers of collisions would never be big enough to make a measurable difference in the pressure at the area investigated.

In a similar way, the total kinetic energy of all the molecules in a given cubic centimeter of our system fluctuates from instant to instant and is not quite the same as the total kinetic energy of those in any other cubic centimeter. Such differences, though, are so small that they defy the ability of thermometers to measure them. Really large fluctuations are so improbable as never to occur. We can thus safely talk about our equilibrium mixture having a specific temperature which is constant with time and uniform throughout.

The behavior of the composition, the pressure, and the temperature of the particular equilibrium situation we have discussed is in fact typical of *all* equilibrium states. The temperature, pressure, and composition are apparently constant with time. In actuality, all three properties are fluctuating *imperceptibly* about a most probable mean. We do not normally notice these fluctuations because they are difficult, and usually impossible, to measure. Any fluctuation which is large enough to be measured is so improbable that it can be ignored.

UPHILL, DOWNHILL, AND PROBABILITY

To regard an equilibrium state as a continuous series of imperceptible fluctuations around a most probable mean is to understand the reason behind the uphill and downhill character of chemical reactions. A downhill reaction, which corresponds to a movement from an out-of-equilibrium to an equilibrium state, can be visualized as a movement from an improbable to a probable situation. If half a mole each of H_2 and D_2 are allowed to react, the laws of probability tell us the inexorable result. Not only is *some* HD formed, but the fraction of HD molecules is 50 ± 0.000000005 per cent, to an accuracy beyond the capability of any analytical technique. Any other distribution outside these limits is improbable beyond imagination.

In a similar way, we must view an uphill reaction as the movement from a probable to an improbable situation. The equilibrium mixture of D_2, H_2, and HD does not proceed to change itself into pure HD. Such a process is not impossible, it is merely an inconceivably improbable fluctuation.

BARRIERS TO THE ATTAINMENT OF EQUILIBRIUM

Although an equilibrium state corresponds to the most probable situation, it must be realized that it is not always possible to attain this state of affairs. The H_2-D_2-HD equilibrium is a case in point. If H_2 and D_2 gases are mixed in the absence of a catalyst, no constant reshuffle of molecular partners occurs. The molecules simply bounce off each other. In order for a reaction to occur, an H-H or a D-D bond must first be ruptured. Such a process requires a great deal of energy. At room temperature, essentially no collisions between H_2 or D_2 molecules are energetic enough for rupture to occur. Thus no molecules can interchange atoms and no HD molecules can be formed. Our arguments about probability are then all in vain. The system remains locked in the H_2 + D_2 situation.

Only if a catalyst is added is a genuine equilibrium state attained. We need not inquire into the precise mode of action of the catalyst at this point. It is sufficient to realize that it allows H_2 and D_2 molecules to adhere to its surface and exchange partners without the requirement of a large amount of energy. Interchange of partners is thus facilitated and proceeds rapidly.

We often encounter systems which appear to be at equilibrium but in fact are not. These situations invariably arise because there is some energy barrier on a molecular level, such as the rupture of a strong bond, preventing a free interchange of atoms between molecules. Without such constant interchange, we cannot apply the laws of probability.

Perhaps the following analogy will make the situation clearer. At least one manufacturer of playing cards always packages his decks with the ace of hearts as the top card. The probability of drawing the ace of hearts from the top of such a deck is *not* 1/52. It is one. Unless the deck is shuffled, the usual laws of probability do not apply. In chemistry, as in poker, it is not always easy to tell when a proper shuffle is taking place.

PROBLEMS 1
(The answers to these and subsequent problems can be found at the end of the book.)

1. What is the probability of drawing the seven of diamonds from a deck of 52 cards?

2. What is the probability of drawing an ace from a deck of cards?

3. What is the probability of drawing a club from a deck of cards?

4. If one die is thrown, what is the probability of a three coming up?

5. If two dice are thrown, what is the probability of threes coming up on both dice?

6. If two dice are thrown, what is the probability that one die will come up six and the other three?

7. If two dice are thrown, what is the probability that a total of nine will come up?

8. A given coin is flipped six times in succession and falls heads each time. What is the probability of this occurring? What is the probability that the coin will fall heads the seventh time it is flipped?

9. One mole of H_2 and one of D_2 are mixed and allowed to attain equilibrium. Two molecules are removed from the equilibrium mixture. What is the probability that the first is H_2 while the second is HD?

10. Assuming that H and D atoms behave identically, deduce the equilibrium constant K_c for the following reactions:

$$H_2O + D_2O \rightleftharpoons 2\ HDO$$

$$NH_3 + ND_3 \rightleftharpoons NH_2D + NHD_2$$

11. Calculate the equilibrium compositions in the two cases given in Problem 10 when one mole of each of the reactants is mixed.

12. Calculate the equilibrium compositions in the two cases given in Problem 10 when one mole of hydrogen compound is mixed with two moles of deuterium compound.

13. A system consists of two molecules and involves two H atoms and two D atoms. If the two molecules are constantly interchanging partners, find the probability of:
    (a) both molecules being HD
    (b) one molecule being H_2 and the other D_2

14. Without consulting Table 1-2, write out all of the seventy arrangements listed there, not necessarily in the same order. Hence, check that there are only 70 possibilities.

15. Argue that the following two processes correspond to the movement from a less probable to a more probable state. (Hint: Argue the low probability of the reverse process occurring.)
    (a) 1 g Hg ($20°C$) + 1 g Hg ($30°C$) $\rightarrow$ 2 g Hg ($25°C$).
    (b) 25 ml 2M NaCl + 25ml H_2O $\rightarrow$ 50ml 1M NaCl solution

Chapter 2

THE POPULATIONS OF ENERGY LEVELS

INTRODUCTION

In Chapter 1 we investigated some of the statistical aspects of the equilibrium

$$H_2 + D_2 \rightleftharpoons 2\,HD \tag{2.1}$$

using the simplifying assumption that H and D atoms behave identically. Such a presupposition is only of use when discussing equilibria in which isotopes are being constantly interchanged. It is of no use whatever in dealing with an equilibrium situation such as

$$H_2 + Cl_2 \rightleftharpoons 2\,HCl \tag{2.2}$$

since by no stretch of the imagination can H and Cl atoms be assumed to behave at all comparably.

As you may already know, this second equilibrium lies well to the right. When H_2 and Cl_2 gases react, the reaction goes virtually to completion, and almost no H_2 and Cl_2 molecules are left once equilibrium has been established. The behavior of equilibrium (2.2) is thus very different from that of equilibrium (2.1). If our study of equilibrium is going to be of any use at all, it should be able to explain such a difference.

Although we are not yet in a position to give such an explanation, we can at least note one difference between the two reactions which may be of some significance to their relative equilibrium behaviors. When 2 moles of HCl are

23

formed from one mole of H_2 and one mole of Cl_2, heat energy is *released*. Such a reaction, in which heat energy is released, is said to be EXOTHERMIC. If reaction (2.2) is exothermic, this means that the system has *lost* energy. In other words, 2 moles of HCl at room temperature have *less* energy than one mole of uncombined H_2 plus one mole of uncombined Cl_2 at the same temperature, the difference being the amount of energy released. By contrast, if H and D atoms behave identically, there should be no such energy difference between products and reactants in reaction (2.1). (Experimentally, there is a barely measurable difference.)

If equilibrium states are to be interpreted in terms of probability, then the comparison between equilibria (2.1) and (2.2) immediately suggests that the *energy* of a molecule has some bearing on the *probability* of its being present in an equilibrium mixture. It is tempting to go even further and suggest that *lower energy* species are *more probable* than higher energy species. Such an assertion, though, carries with it the implication that *only* exothermic reactions are downhill ones, while those which absorb energy (*endothermic* reactions) are uphill, and will not occur on their own. This is not the case at all. There are quite certainly endothermic reactions which can and do occur on their own. Nevertheless, the suggestion has some contact with reality. Although endothermic reactions occur, they are rare. Most downhill reactions are *exothermic*. There should be some merit, therefore, in the suggestion that the energy change in a reaction is a factor influencing the position of an equilibrium. Although it is not the only factor, it is an important one. If we are to investigate chemical equilibrium in terms of probability, we should include energy in our considerations.

The first thing we must appreciate about energy in this connection is that all *energy is quantized*. The student will already be familiar with some examples of quantized energy, particularly in atoms. He knows, for instance, that electrons in an atom can only possess certain fixed energies, and that energies lying between these fixed values are not possible.

The energy which electrons in a molecule can have is similarly quantized, and we can refer to the *electronic energy levels* which an electron or several electrons can occupy in the molecule. A molecule also has other forms of quantized energy deriving from the motion of *nuclei* rather than that of electrons. It is not only the energy of electrons which is quantized. *All* energy is quantized.

In the case of a diatomic gas such as HCl, we can speak of the molecules as having three different kinds of energy in addition to their electronic energies. First, the molecules can move around in their container. The energy which they have from this movement is called their *translational* energy. Second, the molecules can rotate around their centers of gravity, and hence have *rotational* energy. Third, the atoms in a given molecule can move with respect to each other; i.e., they can vibrate. Thus, diatomic molecules can also have *vibrational* energy. All these three kinds of energy, translational, rotational, and vibrational, are *quantized* like electronic energy. Just as we can speak of electrons occupying energy levels, so we can talk about molecules occupying translational energy levels, rotational energy levels, and vibrational energy levels.

We will discuss how these various kinds of energy levels differ from each other in the next chapter. For the moment, in order to be able to explore what connection energy has with probability, we will find it sufficient to know that all energy is quantized.

A SIMPLE EXAMPLE

Our first example of a system in an equilibrium situation in which account is taken of the energy is purposely a simple one. It is a crystal comprising only three identical atoms. We assume that these atoms are free to vibrate about their mean positions in one direction, and that they have no rotational or translational energy. They are thus able to occupy a set of vibrational energy levels. Such a set of energy levels is shown in Figure 2-1

In this figure, the level labeled zero corresponds to the lowest possible vibrational energy. This level of lowest energy is usually called the *ground state* or the *ground level.* The other levels shown, labeled 1, 2, and 3, correspond to higher vibrational energies.

The particular set of levels shown in Figure 2-1 is said to be *evenly spaced* because the energy difference between successive levels is the same. In other words, the energy required to excite an atom from the ground state to level 1 is the same as that required to excite an atom from level 1 to level 2, and so on. (Not all sets of energy levels are equally spaced, as we shall see in the next chapter, but vibrational levels do have this property, at least to a first approximation.)

Let us first consider our crystal to be at the absolute zero of temperature. This temperature corresponds to the crystal having the lowest possible energy. Such a situation is shown in Figure 2-1, where all three atoms (indicated by circles) are in the ground state.

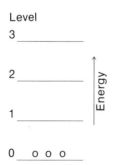

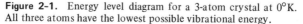

Figure 2-1. Energy level diagram for a 3-atom crystal at 0°K. All three atoms have the lowest possible vibrational energy.

a in Level 1   b in Level 1   c in Level 1

Figure 2-2. If the crystal absorbs one quantum of energy, one of the atoms will be excited. This gives us the three possibilities shown.

As the temperature is raised above absolute zero, the crystal absorbs energy. The atoms are thus excited to vibrate more strongly. They may be excited to level 1 or level 2 or even to higher levels not shown in the figure. Suppose that, at first, just one quantum of energy is absorbed. One of the three atoms in our crystal will absorb this quantum of energy and hence vibrate more strongly than the other two. There are three possible ways in which this can occur, since any one of the three atoms can be excited. These three possibilities are shown in Figure 2-2, where the atoms are labeled a, b, and c. These same three possibilities can also be shown on an energy level diagram as in Figure 2-3.

Suppose now that it is atom b which is excited by the absorption of one quantum of energy. The crystal will not persist in this state with only atom b vibrating more strongly. The quantum of energy will be quickly transferred, perhaps by contact, to atom a or atom c. This other atom will now become vibrationally excited, and atom b will return to the ground state. Such a process will continue indefinitely, the quantum of energy being constantly transferred from one atom to another. Since there is nothing that distinguishes the three atoms from each other, we now assume that it is *equally probable* that atom a, atom b, or atom c is excited. In other words, we assume that three possibilities (i), (ii), and (iii) shown in Figure 2-3 are equally probable.

Now let us consider what happens when the temperature is raised further so that two more quanta of energy are absorbed, giving the system a total of *three* quanta. There are three ways in which the system can accommodate these three quanta:

(A) One atom can have all three quanta.

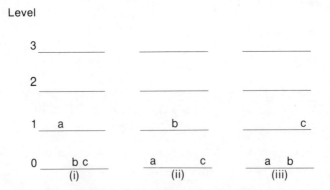

Figure 2-3. The three possibilities of Figure 2-2 shown as energy level diagrams.

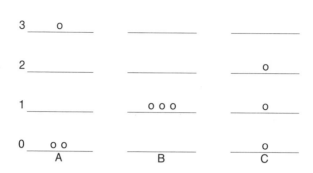

Figure 2-4. If the crystal absorbs three quanta of energy, three different distributions are possible.

(B) Each atom can have one quantum.

(C) One atom can have two quanta, another atom can have one quantum, while the third remains in the ground state.

These three possibilities are shown in Figure 2-4.

Each of the three possibilities in Figure 2-4 is called a *distribution*. To specify a distribution, we need only indicate the *population* of each level; i.e., we need only say *how many* atoms there are in each level. For instance, any situation in which level 3 has one atom and the ground state has two atoms, belongs to distribution A, irrespective of whether atom a, b, or c is excited to level 3.

Once we begin to take account of which atoms are in which level, we find that there is more than one way of achieving some of the distributions in Figure 2-4. Consistent with distribution A, for instance, are *three* different arrangements, depending on whether atom a, b, or c has been excited. These are shown in Figure 2-5 (i), (ii), and (iii). Such different arrangements are called *microstates*.

Similarly there are six ways of achieving distribution C; that is, there are six microstates corresponding to this distribution. These are the arrangements (v) through (x) in the figure. Either a, b, or c an be excited to level 2, leaving two atoms to be accommodated in levels 0 and 1. If a is doubly excited, either b *or* c can be singly excited, leaving the other atom in the ground state. There are thus two microstates ((v) and (vi) in the figure) consistent with atom a being doubly excited. Similarly there are two microstates in which atom b is doubly excited and two more with c, making a total of six microstates.

Finally, there is only one microstate consistent with distribution B. Since *all* three atoms must be singly excited in this distribution, none of the three is *different* from the others. In consequence, only one microstate is possible.

The atoms in the crystal of our example will constantly be transferring energy from one to the other. The system will not be locked in any one of the ten different microstates of Figure 2-5, but will be continually changing from one

DISTRIBUTIONS

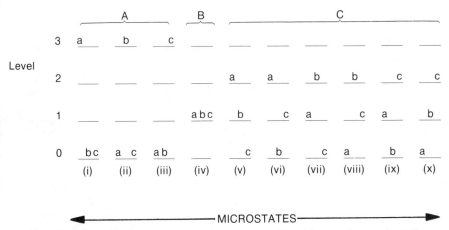

Figure 2-5. The distributions of Figure 2-4 can be achieved in several ways, depending on which atoms are excited.

microstate to the other. Since there is no reason to think otherwise, we now *assume* that *each one* of these ten microstates is *equally probable,* and that the system will be

  in distribution A for 3/10 = 30 per cent of the time,
  in distribution B for 1/10 = 10 per cent of the time, and
  in distribution C for 6/10 = 60 per cent of the time.

Notice that the probability of each distribution is proportional to the corresponding number of microstates. Distribution C is consistent with six different microstates, while A is consistent with three. C is thus twice as probable as A. This behavior is a direct consequence of the assumption that **all microstates are equally probable.**

The number of microstates corresponding to a given distribution is an important quantity. It is called the *thermodynamic probability* and given the symbol W. Thus

$$W = 3 \text{ for distribution A}$$

$$W = 1 \text{ for distribution B}$$

$$W = 6 \text{ for distribution C}$$

Although W is called the thermodynamic probability, strictly speaking it is not a probability in terms of the definition given in Chapter 1, since it is not a fraction and the sum of all the W's does not equal unity. Nevertheless, the thermodynamic probability of a distribution is *proportional* to its true probability.

The approach we have used for a system of three atoms, equally spaced levels, and three quanta of energy, can, of course, be used for *any* system, i.e., for any number of particles, for energy levels which are not necessarily equally spaced, and for any amount of energy. No matter how large or how small the system, we can always consider it in terms of distributions, microstates, and thermodynamic probabilities. We can assume for *any* system as we did for our simple system that *all microstates corresponding to the same total energy are equally probable.* If W is the number of microstates consistent with a given distribution, then the probability of that distribution is directly proportional to W.

PROBLEMS 2.1
(The problems in this chapter should be attempted whenever they are encountered in the body of the text.)

1. A crystal consists of four atoms with equally spaced vibrational energy levels. If such a system has 3 quanta of energy, only three distributions are possible. These are given in Figure 2-6.

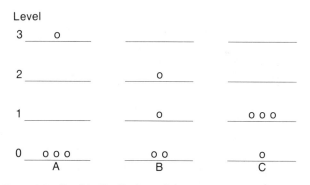

Figure 2-6. Possible distributions of three quanta among four atoms.

Show that W = 4 for distribution A, W = 12 for B, and W = 4 for C.

2. If the system described in (1) has only 2 quanta of energy, show that only two distributions are possible, one with W = 6, and the other with W = 4.

PERMUTATIONS AND FACTORIALS

In order to go further with our discussion of the probabilities of various distributions of atoms or molecules in energy levels, we must find a general formula for calculating W. Without such a formula we are restricted to rather simple systems involving only a handful of molecules.

Let us first consider the problem of how many ways there are of arranging a number of objects in a row. Such arrangements are usually called *permutations.*

It is quite easy to deduce the number of permutations in the general case, i.e., for any number of objects. The trick is to start with the case of two objects and move up through three and four objects to any number.

With two objects, a and b, there are two arrangements, ab and ba. The second object, b, can be added to the first in two ways: either in front of a or behind it. The number of permutations of three objects is now easily deduced. There are three ways of adding the third object to either ab or ba: at the beginning, in between the two objects, or at the end. This gives us, in the case of ab, the three permutations cab, acb, and abc; and in the case of ba, the three permutations cba, bca, and bac. We thus have a total of six permutations. This number 6 is best thought of as being the product $3 \times 2 \times 1$. The 1 corresponds to the one arrangement of a alone. The 2 corresponds to the *two* ways of adding the *second* object to a, while the 3 corresponds to the three ways of adding the *third* object to any given permutation of the first two.

By looking at the problem in this fashion, we can easily go on to deduce the number of permutations of any number of objects. The number of permutations of four objects, for instance, is obviously $4 \times 3 \times 2 \times 1$, since there are *four* ways of adding a fourth object to any permutation of three objects. These four positions are shown in Figure 2-7.

Since there are already $3 \times 2 \times 1$ permutations of the three objects, this means a total of $4 \times 3 \times 2 \times 1$ permutations of the four objects. Extending this argument to five objects, we deduce the number of permutations as $5 \times 4 \times 3 \times 2 \times 1$. Clearly, a formula for the number of permutations of n objects must be:

$$n(n-1)(n-2)(n-3)(n-4) \ldots \times 3 \times 2 \times 1.$$

If n is an integer, then the number resulting from the repeated multiplication $n(n - 1)(n - 2)(n - 3) \ldots \times 3 \times 2 \times 1$ is given a special name. It is called *"factorial n"* or *"n factorial"* and is indicated, a little surprisingly, by the symbol

Figure 2-7.   Given three objects in a row, there are four ways of adding a fourth object to the row.

Level

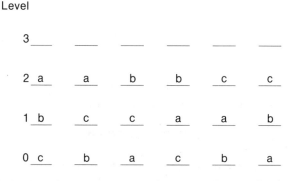

Figure 2-8. The six ways of achieving distribution C of Figure 2-4.

n!. Thus 4! = 4 × 3 × 2 × 1 = 24; 5! = 5 × 4 × 3 × 2 × 1 = 120; and so on. *The number of permutations of n objects is thus* n!.

The problem of finding how many ways there are of arranging n objects in a row, which we have just solved, is really the same problem as finding the number of ways of achieving a distribution in which there is *one atom per energy level.* Consider, for instance, distribution C of Figure 2-4, in which there is one atom in each of three energy levels. In Figure 2-8 the six ways of achieving this distribution are illustrated. We could write these six possibilities down as abc, acb, bca, bac, cab, and cba to show that they correspond to the six permutations of three objects. In the diagram, level 2 then corresponds to "first place" in the sequence, level 1 to second place, and level 0 to third place. The problem is obviously the same in both cases, not only for three objects and three levels but for n objects and n levels. The number of ways of achieving a distribution in which n particles each occupy one of n levels is n!.

PROBLEMS 2.2

   1. Evaluate 7! and 8!.

   2. Is 13! a factor of 43!?

   3. Under what circumstances is x! a factor of y!?

   4. Evaluate 8!/7!.

   5. Evaluate n!/(n −1)!.

   6. You are given two decks of cards, both recently shuffled. What is the probability that if you shuffle one deck, it will turn out to have the cards in the same order as in the other deck?

_____		_____	_____	_____	_____

o		d	c	b	a

o o o		a b c	a b d	a c d	b c d
		(i)	(ii)	(iii)	(iv)

Figure 2-9. A simple distribution and the four microstates corresponding to it.

A FORMULA FOR W

We now turn to the problem of finding W for distributions in which there is *more than one atom* in some or all levels. Consider, for instance, the distribution shown in Figure 2-9. In this distribution there are three atoms in the ground state and one in an excited state.

There are obviously only four microstates corresponding to this distribution so that W = 4. Any one of the four atoms can be the excited one.

We can arrive at this same answer by a different route using factorials. The number of permutations of the four atoms a, b, c, and d is of course 4! = 24. W is less than this number by a factor 24/4 = 6 = 3!. The occurrence of this second factorial is not a coincidence. In taking permutations we would count all the 3! arrangements of Figure 2–10 as different.

Yet, in fact, all six arrangements are really the *same* microstate. They all correspond to atom d being excited and the other three atoms being in the ground state. To say, "Atoms a, b, and c are in the ground state," means *exactly* the same thing as saying, "Atoms c, b, and a are in the ground state." In counting permutations, we count microstate (i) of Figure 2-9 over and over again 3! different ways. The same applies to the other microstates. The number of permutations is therefore 3! times bigger than W, or W = 4!/3! = 4 as previously argued.

Consider next the distribution shown in Figure 2-11. In this distribution there are three atoms in the ground state and two in the first excited state: a total of five atoms. From our previous example we know that 5! is too large a figure for W. Counting permutations considers the 12 arrangements shown in Figure 2-12 to be different, whereas in fact they are all the same microstate. Every microstate is counted 3! times too often because of the 3 atoms in the ground state, and a further 2! times too often because of the two atoms in the first excited state, or a total of 3! × 2! times too many. Thus, W = 5!/3!2! = 10. All ten microstates are shown in Figure 2-13. Exactly the same argument enables

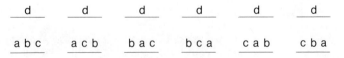

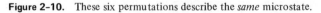

Figure 2-10. These six permutations describe the *same* microstate.

Level

2 _____

1 o o

0 o o o

Figure 2-11. A distribution in which five atoms occupy two levels.

us to arrive at a general formula for W. If there are a total of n atoms, m of which are in one level and (m - n) in a second level, the total number of microstates is given by:

$$W = \frac{n!}{m!\,(m - n)!}$$

Level

2 ___   ___   ___   ___   ___   ___   ___   ___   ___   ___   ___   ___

1 ab   ab   ab   ab   ab   ab   ba   ba   ba   ba   ba   ba

0 cde   ced   dce   dec   ecd   edc   cde   ced   dce   dec   ecd   edc

Figure 2-12. These twelve permutations are not different; they describe the *same* microstate.

We need not stop at cases where only two levels are occupied. Our argument is valid for any number of occupied levels. If n is the total number of molecules, and n_0 the population of the ground state, n_1 the population of level 1, n_2 that

Level

2___   ___   ___   ___   ___   ___   ___   ___

1 ab   ac   ad   ae   bc   bd   be   cd   ce   de

0 cde   bde   bce   bcd   ade   ace   acd   abe   abd   abc

Figure 2-13. The ten microstates of the distribution of Figure 2-11.

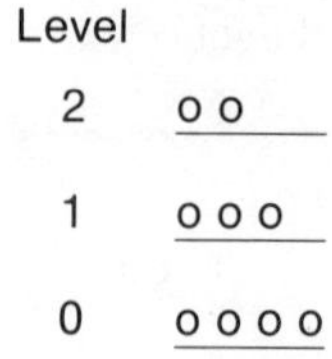

Figure 2-14. A distribution in which nine atoms occupy three energy levels.

of level 2 and so on, then the number of microstates is given by the formula

$$W = \frac{n!}{n_0! n_1! n_2! n_3! \ldots} \qquad (2.3)$$

Thus, for example, W for the distribution shown in Figure 2-14 is given by $W = 9!/4!3!2! = 1260$. We would have to spend a long time counting out all these possibilities if we did not have a formula to do it for us. Nevertheless, the system in Figure 2-14 still contains only a few molecules. If we were to consider a mole of molecules (6×10^{23}), W would be inconceivably large.

PROBLEMS 2.3

1. A system consists of 7 atoms which can occupy equally spaced levels. Four quanta of energy are added to the system. Two out of five possible distributions are shown in Figure 2-15. Show that $W_A = 7$ and $W_B = 21$. Also write down the three remaining distributions and calculate their thermodynamic probabilities. Which is the most probable of the five distributions?

Distribution A Population	Level	Distribution B Population
1	4	—
—	3	—
—	2	2
—	1	—
6	0	5

Figure 2-15. Two possible distributions of seven atoms having four quanta of energy.

2. A system consists of 15 atoms which can occupy equally spaced energy levels. Eleven quanta of energy are supplied. It can be shown that the most probable distribution for this situation is:

Level	Population
3	1
2	2
1	4
0	8

By trial and error, find the second most probable distribution, and show that in this second distribution the populations of each level are not very different from those given above. Show also, with a few examples, that distributions in which levels 1 and 2 are both unpopulated are much less probable than the given distribution. Make sure that all distributions discussed correspond to 15 atoms and 11 quanta.

3. The number of arrangements given in Table 1-2 can be calculated from the following argument. We have eight positions, L1, R1, L2, R2, L3, R3, L4, and R4, four of which can be H and four of which can be D. This is mathematically the same problem as finding how many ways there are of putting eight atoms into two levels, with four atoms per level; i.e., find W for the distribution:

oooo
———

oooo
———

Show that this argument gives the correct number of arrangements shown in Table 1-2, namely 70.

SOME MORE COMPLEX SYSTEMS

Now that we have a general formula for W we can begin to deal with systems in which there are a large number of molecules or atoms. Consider first the addition of 4 quanta of energy to a system consisting of a large number of atoms which can occupy equally spaced energy levels. Calling the total number of atoms n, we have the five distributions shown in Figure 2-16. No other distributions are possible. Application of formula (2.3) to distribution A gives

$$W_A = \frac{n!}{(n-1)!} = \frac{n\,(n-1)(n-2)(n-3)\dots 1}{(n-1)(n-2)(n-3)\dots 1} = n.$$

Very conveniently, all but one of the terms in n! cancel with those in $(n-1)!$.

Level			Populations		
4	1	—	—	—	—
3	—	1	—	—	—
2	—	—	2	1	—
1	—	1	—	2	4
0	$(n-1)$	$(n-2)$	$(n-2)$	$(n-3)$	$(n-4)$
Distribution					
	A	B	C	D	E

Figure 2-16. The five distributions which are possible when four quanta of energy are supplied to a system having a large number of molecules.

Similarly we find that:

$$W_B = \frac{n!}{(n-2)!} = \frac{n(n-1)(n-2)(n-3)\ldots}{(n-2)(n-3)\ldots} = n(n-1)$$

Now, if n is a very large number comparable, say, to the Avogadro number, we can make the approximation $(n-1) \approx n$, giving $W_B \approx n^2$.

Similar arguments lead to the other values of W:

$$W_C = \frac{n!}{(n-2)!2!} = \frac{n(n-1)}{2} \approx \frac{n^2}{2}$$

$$W_D = \frac{n!}{(n-3)!2!} = \frac{n(n-1)(n-2)}{2} \approx \frac{n^3}{2}$$

$$W_E = \frac{n!}{(n-4)!4!} = \frac{n(n-1)(n-2)(n-3)}{24} \approx \frac{n^4}{24}$$

Of these five thermodynamic probabilities, the last, W_E, is the largest. In fact, it is $n/12$ times as large as W_D, the next largest value of W. If n is of the order of 10^{23} or 10^{24}, as would be the case for a real system, then W_E will be some 10^{22} or 10^{23} times as large as W_D. We can now invoke our basic assumption that all microstates corresponding to the same total energy are *equally probable*. As argued above, this means that the probability of finding the system in any distribution is proportional to the thermodynamic probability W of that distribution. Since distribution E has a much larger value of W than any of the others, it will be very much more probable. The system will *almost always* be in distribution E; since W_E is some 10^{22} times larger than its closest competitor, W_D, distribution E is some 10^{22} times more probable than its closest rival, D. For all practical purposes we can neglect all distributions but E, since the system will almost never be in any other distribution.

It would be very convenient if the most probable distribution always behaved in this fashion, if it were always overwhelmingly more probable than

Level	Population
3	$n_3 = 0$
2	$n_2 = 4.05 \times 10^6$
1	$n_1 = 1.57 \times 10^{15}$
0	$n_0 = 6.02 \times 10^{23}$

Figure 2-17. The most probable distribution of a mole of hydrogen molecules among vibrational levels at $300°K$.

any other. Unfortunately, the situation is not as simple as this. Only in cases where we have a relatively small number of energy quanta do we find that the most probable distribution has no serious competitors. If, as is ordinarily the case, enough energy is available to excite an appreciable number of atoms or molecules into several energy levels above the ground state, the most probable distribution loses its predominance (but not its significance).

As an example to illustrate this point, let us next consider the most probable distribution shown in Figure 2-17. The system illustrated is a mole of H_2 molecules at $300°K$. Only the equally spaced vibrational levels are considered. As can be seen from the figure, a large number of molecules are vibrationally excited at this temperature, not only to the first excited level, but to the second as well.

Although the distribution shown in Figure 2-17 is the most probable distribution for the given temperature, it is only one among a large number of distributions of almost equal probability. Some of these distributions are shown in Figure 2-18.

Here, A is the most probable distribution, i.e., the same one that appears in Figure 2-17. Distribution B can be obtained from A by exciting one molecule from level 1 to level 2, and de-exciting one molecule from level 1 to the ground state. Since one of the shifted molecules has gained a quantum of energy, and the other has lost one, the energies of both A and B are the same. Distribution C can be obtained from A in much the same way as for B. The difference is that C

Level					
3					
2	n_2	$n_2 + 1$	$n_2 - 1$	$n_2 + 2$	$n_2 - 2$
1	n_1	$n_1 - 2$	$n_1 + 2$	$n_1 - 4$	$n_1 + 4$
0	n_0	$n_0 + 1$	$n_0 - 1$	$n_0 + 2$	$n_0 - 2$
	A	B	C	D	E

Figure 2-18. The most probable distribution of Figure 2-17 and some other distributions which are almost as probable.

has two more rather than two less molecules in level 1. Again, since one molecule has been excited and another de-excited, the energy remains unaltered. Distributions D and E are similar to B and C except that twice as many molecules have been shifted into or out of the three occupied levels without altering the total energy.

Both distributions B and C are very nearly as probable as distribution A (see Problems 2.4). They are less probable by only a minute amount, one part in four million. It can be shown, for example, that

$$\frac{W_A}{W_B} = 1 + \frac{1}{4 \times 10^6}.$$

Similarly, distributions C, D, and E are also very nearly as probable as A. Nor are these five distributions the only ones of comparable probability. We can, as it happens, go on for some time forming new distributions from distribution A by pulling molecules out of level 1 two at a time and placing one of them in level 2 and one of them in the ground level, without encountering distributions which are much less probable than distribution A. When we have removed two thousand molecules from level 1 in this way, for instance, we achieve distribution F, shown in Figure 2-19. Distribution F is about 90 per cent as probable as A.

Beyond a figure of 2×10^3 molecules removed from (or added to) level 1, the value of W for the distribution drops more rapidly, and we obtain much less probable distributions. By the time we have shifted 2×10^4 molecules from level 1, as in distribution G, we have a distribution which is only 2×10^{-5} as probable as A.

We should notice that distributions in which level 3 is occupied are moderately improbable. The most probable of such distributions is that labeled H in Figure 2-19. H has about one hundredth the probability of distribution A.

The figures we have quoted show that there are a large number of distributions which are almost as probable as the most probable distribution. This state of affairs can only mean that the most probable distribution does not occur very often. This need not worry us unduly. Such a situation is very much like that

Level			
3			1
2	$n_2 + 10^3$	$n_2 + 10^4$	$n_2 - 2$
1	$n_1 - (2 \times 10^3)$	$n_1 - (2 \times 10^4)$	$n_1 + 1$
0	$n_0 + 10^3$	$n_0 + 10^4$	n_0
	F	G	H

Figure 2-19. Some distributions which are less probable than those shown in Figure 2-18.

encountered in Chapter 1 when we discussed the H_2-D_2-HD equilibrium. There we found that the most probable number of HD molecules in the mixture (exactly 50 per cent of the total) did not occur very often. Despite this fact, it was still meaningful to talk about an equilibrium concentration of HD. In the case under discussion, although it is true that the system is very seldom in its most probable distribution, *it is also true that it is almost never in a distribution which is very different from the most probable one.* It makes sense, therefore, to talk about the most probable distribution shown in Figure 2-17. We do not imply that the system is perpetually locked in this single distribution. Rather, we mean that the system is fluctuating incessantly about this distribution as a mean, never diverging from it sufficiently to make a measurable difference in bulk properties.

What we have just said is true not only of the system of Figure 2-17 but of *all* systems which are in equilibrium with their surroundings. In any such system the atoms or molecules are almost never in a distribution significantly different from the most probable distribution. In the future we will describe such a situation by saying that the atoms or the molecules are in their most probable distribution, even though such a statement, taken literally, is untrue.

PROBLEMS 2.4

1. A system similar to that of Figure 2-16 consists of n particles which can occupy equally spaced energy levels. Three quanta of energy are supplied to the system. Draw diagrams showing the three possible distributions. Calculate W for all three distributions and hence show that one of the distributions is overwhelmingly more probable than the other two.

2. Check that the population figures given in Figure 2-17 form a geometric progression; i.e., show that the ratio n_1/n_0 is the same as the ratio n_2/n_1.

3. Show that for the distributions shown in Figure 2-18

$$W_A/W_B = \frac{(n_2 + 1)(n_0 + 1)}{n_1(n_1 - 1)}$$

and

$$W_B/W_C = \frac{(n_1 + 2)(n_1 + 1)}{n_2 n_0}.$$

Hence show that if $n_1/n_0 = n_2/n_1$, as discovered in Problem 2, then W_A is larger than W_B or W_C.

4. Check that distribution H in Figure 2-19 has about one hundredth of the probability of distribution A (Figure 2-18), by finding the ratio W_A/W_H.

THE BOLTZMANN DISTRIBUTION LAW

Up to the present, we have talked about the most probable distribution of particles among energy levels without inquiring too closely as to the exact form of such a distribution or how we might determine it. Such calculations can be made with the aid of a formula called the *Boltzmann distribution law*. Unfortunately, it is beyond the scope of this book to derive this law from first principles. Students who have worked conscientiously through the problems immediately above (Problems 2.4), will have encountered some of the steps in such a proof.

There are two features of the proof of the Boltzmann law which make it unsuitable for discussion here. The first is the level of mathematics required, which is more than most beginning students will possess. The second is that a long argument is required to justify a scale of temperature. It has been said of temperature that the more you think about it, the less you understand it. In this book it will be assumed that the reader is still in a happy state of ignorance about the concept of temperature. and that he thinks he understands it.

Before turning to a statement of the Boltzmann law, we must introduce two constants which are featured in it. The first is a number which, like π, is so often encountered in mathematics that it is given a symbol of its own. It is the number e. This number is the sum of the series $1 + 1/1! + 1/2! + 1/3! + 1/4! + \ldots$ The reader can easily verify for himself that e has the value $2.71828 \ldots$ We often encounter logarithms which have the number e rather than 10 as a base. These logarithms are written $\log_e x$ or more simply $\ln x$. They are called *natural logarithms*.

The second constant featured in the Boltzmann law is the *Boltzmann constant* k. This constant is closely related to two other constants with which you are already familiar, the gas constant R, and the Avogadro number N. In fact,

$$k = \frac{R}{N} \qquad\qquad (2.4)$$

Since k is equal to the gas constant divided by the Avogadro number, we can think of k as being the gas constant per molecule, while R is the gas constant per mole. The Boltzmann constant has the dimensions of energy per degree per molecule. If the energy is expressed in calories, k has the value 3.2999×10^{-24} cal deg^{-1} $molecule^{-1}$. If ergs are used, k has the value 1.3581×10^{-16} erg deg^{-1} $molecule^{-1}$.

Figure 2-20 illustrates the general case of a distribution of particles over a set of energy levels. Here, n_0, n_1, n_2, n_3, and n_4 are the populations of levels 0, 1, 2, 3, and 4 respectively. ϵ_1 is the energy of level 1, ϵ_2 that of level 2, ϵ_3 that of level 3, and ϵ_4 that of level 4. The energy of the ground state, ϵ_0, is taken to be zero. Thus, for instance, it requires an energy ϵ_1 to excite one molecule from the ground state to level 1.

Level

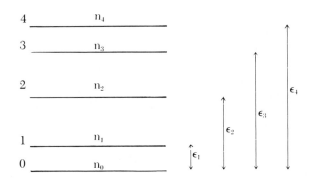

Figure 2-20. A general labelling system for a Boltzmann distribution.

The Boltzmann distribution law may now be stated. It is given by the equation:

$$\frac{n_i}{n_0} = e^{-\epsilon_i/kT} \qquad (2.5)$$

where T is the absolute temperature, ϵ_i is the energy of the ith level, and n_i is its population, while n_0 is the population of the ground state.

We can also express the Boltzmann law in a slightly different form. If the energy of the ground state, ϵ_0, is not taken as the arbitrary zero of energy, then equation (2.5) can be written as

$$\frac{n_i}{n_0} = e^{-\Delta\epsilon_i/kT} \qquad (2.5a)$$

where $\Delta\epsilon_i = \epsilon_i - \epsilon_0$.

In order to become familiar with this law, we will now consider some examples of its use. We start by calculating the distribution, shown in Figure 2-17, of a mole of H_2 molecules over its vibrational levels at $300°$ K. We will make the approximate assumption that the vibrational levels are equally spaced at intervals of 1.956×10^{-20} cal molecule^{-1}. Thus $\epsilon_1 = 1.956 \times 10^{-20}$ cal molecule^{-1} and $\epsilon_2 = 3.912 \times 10^{-20}$ cal molecule^{-1}.

We then have

$$\frac{\epsilon_1}{kT} = \frac{1.956 \times 10^{-20}}{3.299 \times 10^{-24} \times 300} \frac{\text{cal molecule}^{-1}}{\text{cal deg}^{-1} \text{ molecule}^{-1} \times \text{deg}} = 19.77,$$

(Note that the term ϵ_i/kT is *always* a dimensionless, pure number.)

and similarly

$$\frac{\epsilon_2}{kT} = 39.54.$$

We must next evaluate the two numbers $e^{-19.77}$ and $e^{-39.54}$. This is best done by taking logarithms.

If we let

$$x = e^{-19.77},$$

then

$$\log_{10}x = \log_{10}e^{-19.77} = -19.77 \log_{10}e \qquad \text{(since log } a^b = b \log a)$$

Now

$$\log_{10}e = 0.4343 \qquad \text{(a number worth remembering)}$$

so that

$$\log_{10}x = -19.77 \times 0.4343$$

$$= -8.586$$

$$= 0.414 - 9$$

Hence

$$x = \text{antilog}\,(0.414 - 9) = 2.59 \times 10^{-9}.$$

By an exactly similar process,

$$e^{-39.54} = 6.73 \times 10^{-18}$$

We thus have from the Boltzmann law that:

$$\frac{n_1}{n_0} = e^{-\epsilon_1/kT} = e^{-19.77} = 2.59 \times 10^{-9} \qquad \textbf{(2.6)}$$

and

$$\frac{n_2}{n_0} = e^{-\epsilon_2/kT} = e^{-39.54} = 6.73 \times 10^{-18} \qquad (2.7)$$

Now, the total number of molecules is 6.023×10^{23}, of which only a very tiny fraction, given in (2.6) and (2.7), are excited vibrationally. We can thus approximate by making

$$n_0 = 6.023 \times 10^{23},$$

so that, from (2.6),

$$n_1 = 6.023 \times 10^{23} \times 2.59 \times 10^{-9}$$

$$= 1.56 \times 10^{15},$$

while from (2.7),

$$n_2 = 6.023 \times 10^{23} \times 6.73 \times 10^{-18}$$

$$= 4.05 \times 10^{6}.$$

In the example just calculated, we expressed the energies of the levels in terms of energy per *molecule*. It is often more convenient to express such energies in *molar* rather than molecular amounts. Instead of describing the levels in our example as being 1.956×10^{-20} cal molecule^{-1} apart, we could have described them as being 11.79 kcal mole^{-1} apart. In other words, it requires 11.79 kcal of energy to excite a whole mole of H_2 molecules from the ground state to the first excited state. (It is easily checked that 1.956×10^{-20} cal multiplied by the Avogadro number gives 11.79 kcal.)

The expression for the Boltzmann law, equation (2.5), can easily be modified to deal with cases in which energies are given in molar rather than molecular amounts. If ϵ_i is the energy of a level *per molecule*, while E_i is its energy *per mole*, then

$$\epsilon_i = \frac{E_i}{N},$$

where N is the Avogadro number. Thus,

$$\frac{\epsilon_i}{kT} = \frac{E_i}{NkT} = \frac{E_i}{RT}$$

Feeding this back into equation (2.5), we have our final result:

$$\frac{n_i}{n_0} = e^{-E_i/RT}.$$

(2.8)

As an example of the use of equation (2.8), we will calculate the population of the first excited vibrational level (level 1) for a mole of HCl gas molecules at 25°C, given that this level is 8.25 kcal mole^{-1} higher in energy than the ground state.

Since $R = 1.987$ cal deg^{-1} mole^{-1} and $T = 298.2°K$,

$$\frac{E_1}{RT} = \frac{8.250 \times 10^3 \text{ cal mole}^{-1}}{(1.987 \times 298.2) \text{ cal mole}^{-1}} = 13.92$$

From equation (2.8) we then find that

$$\frac{n_1}{n_0} = e^{-13.92}.$$

In the same way as before, let

$$x = e^{-13.92}.$$

Then

$$\log_{10} x = -13.92 \log_{10} e = -6.044$$

so that

$$x = \text{antilog}(-6.044) = \text{antilog}(0.956 - 7)$$

$$= 9.04 \times 10^{-7}.$$

Again we make the approximation:

$$n_0 = N = 6.023 \times 10^{23}.$$

Thus

$$n_1 = 6.023 \times 10^{23} \times 9.04 \times 10^{-7},$$

giving the final result:

$$n_1 = 5.44 \times 10^{17}.$$

SOME FEATURES OF A BOLTZMANN DISTRIBUTION

Since the Boltzmann law involves the factor $e^{-\epsilon_i/kT}$, it is just as well to familiarize ourselves with the behavior of this factor, and in particular how it varies with both ϵ_i and T. In order to do this, let us call

$$x = \frac{E_i}{RT} = \frac{\epsilon_i}{kT}. \qquad (2.9)$$

The Boltzmann law, equation (2.5), then becomes

$$\frac{n_i}{n_0} = e^{-x}. \qquad (2.10)$$

In Figure 2-21, the way in which e^{-x} varies with x is illustrated. For x = 0, the value of e^{-x} is 1. As x increases, e^{-x} decreases. For instance, when x = 2, e^{-x} = 0.135; by the time x = 5, e^{-x} is only 0.0067. As x becomes larger and larger, e^{-x} approaches zero as a limit.

Now, since $x = E_i/RT$, the smaller T is, the larger x will be. We thus expect that at low temperatures x will be large, and that e^{-x} and hence n_i/n_0 will be very small. In other words, the closer we come to absolute zero, the smaller n_i will be relative to n_0. In the limit, only the ground state will be occupied at $0°K$.

Conversely, the higher the temperature, the smaller x becomes, and hence the larger e^{-x} becomes. As T approaches infinity, x approaches zero, and e^{-x} approaches one. Thus, with increasing temperature, n_i/n_0 increases, until at a sufficiently high temperature it approaches, but never quite reaches, unity. No single level is ever as populated as the ground state, no matter what the temperature.*

In Figure 2-22 we illustrate how the Boltzmann distribution varies with temperature for a set of evenly spaced levels. The system involved is the same as in Figure 2-17. The levels are 11.79 kcal mole^{-1} apart, and approximately describe the vibrational levels of H_2 molecules. The figure refers to a mole of H_2 molecules. Notice how, as predicted, only the lower levels are populated at low temperatures, and how, with increasing temperature, the population of a particular level (level 1, for example) becomes closer to that of the ground state.

The value of x in equations (2.9) and (2.10) depends not only on the temperature T but also on the energy ϵ_i. The larger the energy, the larger x is, and the smaller e^{-x} is. We should thus expect, as illustrated in Figure 2-22, that the higher the energy of a level, the smaller its population, and vice versa.

---

*Sometimes, though, we have the phenomenon of degeneracy. If a level is *degenerate,* there are several sublevels of equal energy. In the hydrogen atom, for instance, p-levels all consist of three levels of equal energy. They are said to be triply degenerate. If excited levels are degenerate, but the ground state is not, it sometimes happens that although each of these degenerate levels has a population less than n_0, the collective population of all levels taken together is larger than n_0. Such behavior is found, for instance, in the case of distributions among rotational levels.

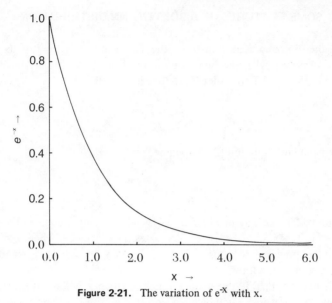

Figure 2-21. The variation of e^{-x} with x.

There is one further feature of a Boltzmann distribution which should be noted, namely that the population of a given level depends not only on the energy and the temperature, but also on the existence or non-existence of other levels. This point is best made by an example. In Figure 2-23 two Boltzmann distributions of a mole of particles over two different sets of energy levels are

Level				
10				
9				4
8				1.51×10^{4}
7				5.67×10^{6}
6			10	2.13×10^{9}
5			9.33×10^{4}	8.01×10^{11}
4			8.55×10^{8}	3.01×10^{13}
3		2.94×10^{4}	7.83×10^{12}	1.13×10^{16}
2		8.05×10^{11}	7.17×10^{16}	4.25×10^{18}
1	4.07×10^{6}	2.20×10^{17}	6.57×10^{20}	1.60×10^{21}
0	6.02×10^{23}	6.02×10^{23}	6.02×10^{23}	6.01×10^{23}
	150°K	400°K	650°K	1000°K

Figure 2-22. Boltzmann distribution of a mole of hydrogen atoms among the vibrational levels at four different temperatures.

Level		Energy	Level	
10	1.44×10^{21}	3000	1	3.78×10^{21}
9	2.40×10^{21}	2700		
8	4.01×10^{21}	2400		
7	6.68×10^{21}	2100		
6	1.11×10^{22}	1800		
5	1.86×10^{22}	1500		
4	3.10×10^{22}	1200		
3	5.16×10^{22}	900		
2	8.61×10^{22}	600		
1	1.44×10^{23}	300		
0	2.39×10^{23}	000	0	6.02×10^{23}

(center column labeled: calories per mole)

Set A   Energy   Set B

Figure 2-23. Two distributions having different energy level spacings. Both are at the same temperature. Both refer to a mole of particles.

illustrated. Both are at the same temperature. In one case, labeled A, the energy levels are spaced at a distance of 300 cal mole^{-1} from each other. In the other case, labeled B, the levels are ten times more widely spaced, the difference between successive levels being 3 kcal mole^{-1}. Now compare the populations of level 10 in A with level 1 in B. Although the energies of these two levels are the same, and the temperature is the same, the level in the more widely spaced set has a higher population. This is so despite the fact that the ratio n_{10}/n_0 in set A and the ratio n_1/n_0 in set B are identical. In set A, both ground state and level 10 have less population than the corresponding levels of the same energy in set B, because of the competition for population from the intervening levels. The population of a given energy level thus depends not only on its energy and on the temperature, but also on whether there are other levels in the set competing with it for population. In general, the closer energy levels are to each other, i.e., the more closely spaced they are, the more severe such a competition will be. As we shall see presently, the spacing of energy levels is not without its influence on chemical equilibrium.

PROBLEMS 2.5

1. The vibrational energy levels of HCl molecules are approximately evenly spaced and about 8.25 kcal mole^{-1} apart. Calculate the populations of the lowest three levels in a mole of this gas at $100°$ K, $200°$ K, and $300°$ K.

2. Calculate the ratio n_1/n_0 for the two lowest vibrational levels at $25°$ C for the following molecules.

Molecule	Energy difference between the two lowest vibrational levels
H_2	11.89 kcal mole^{-1}
D_2	8.55
O_2	4.45
F_2	2.54
I_2	0.610

3. The two lowest rotational levels of Cl_2 are 1.38 cal mole^{-1} apart, while the two lowest vibrational levels are 1.59 kcal mole^{-1} apart. By calculating n_1/n_0 for the two cases at $25°$ C, show that a very much larger percentage of molecules is rotationally excited than is vibrationally excited.

4. Calculate the ratio n_1/n_0 for the two lowest rotational levels at $25°$ C for the following molecules:

Molecule	Energy difference between the two lowest rotational levels
H_2	339 cal mole^{-1}
D_2	169
O_2	11.5
Br_2	0.463

5. Find a table of values for e^{-x}. Use it to plot e^{-x} against x over the range x = 0 through x = 1.0.

6. Calculate the temperatures of the two systems in Figure 2-23.

7. To what temperature must H_2 gas be heated for one molecule in a thousand to be vibrationally excited (assume the levels to be evenly spaced at intervals of 11.8 kcal mole^{-1})?

8. Cross-check the population figures for $650°$ K in Figure 2-22.

9. N particles are distributed over an equally spaced set of energy levels at temperature T. The energy difference between the levels is E cal mole^{-1}. Show that the population of the ground state is given by the equation

$$n_0 = N (1 - e^{-E/RT}).$$

(A knowledge of geometric series is required for this problem.)

10. A mole of particles is distributed over a set of energy levels which are equally spaced at intervals of 60 cal mole^{-1}. Use the equation given in the previous question to calculate the population of the ground state for this system at 25°C.

Chapter 3

THE
SPACING
OF
ENERGY
LEVELS

INTRODUCTION

In the previous chapter, we took the existence of energy levels more or less for granted. The mere fact that energy levels exist was enough to allow us to argue about distributions of particles over such levels. The exact nature and origin of the levels did not concern us overly much. In this chapter, by contrast, it is precisely such aspects of energy levels which we shall discuss. In particular, we will investigate the nature of the four different kinds of energy levels we commonly encounter, namely translational, rotational, vibrational, and electronic levels. What will be of most interest to us about these levels is their spacing. We will want to know whether the levels are relatively close to one another in energy or far apart. We have already seen that the spacing of energy levels is one of the factors determining their populations. When we re-examine chemical equilibrium in terms of energy levels, we will see that the spacing of levels plays a very important role in determining the position of equilibrium.

The chief difficulty in discussing the spacing of energy levels is the mathematical complexity of the subject. In order to calculate the exact value of one of the quantized energies which a molecule or an atom can possess, we must first solve the Schrödinger wave equation. Even in the simplest case, such a solution is entirely beyond the mathematical level set for this text. This forces us to use a simpler, approximate approach to the problem.

The approach we shall use here is one first put forward by the French physicist Count Louis de Broglie a few years before the appearance of Schrödinger's equation in 1926. It could be said that de Broglie provided the seed from which Schrödinger's equation soon grew.

De Broglie made the original suggestion that what we commonly regard as moving particles behave in some respects like waves, and that it is this wave nature of particles which accounts for the quantization of their energies. He proposed that if a particle has a mass m and a velocity v, then it also has a wave associated with it of wavelength λ given by the formula:

$$\lambda = \frac{h}{mv} \tag{3.1}$$

where h is a physical constant called Planck's constant.

How such an idea can be used to predict quantized energy levels is perhaps best seen by means of an example, the so-called "particle in a one-dimensional box."

A PARTICLE IN A ONE-DIMENSIONAL BOX

Imagine a particle, such as that shown in Figure 3-1, which is able to move in one direction only, the x-direction. The particle is confined by two impenetrable walls, perpendicular to the x-direction, a distance d apart. The left-hand wall is considered to be at x = 0 and the right-hand wall at x = d.

If the mass of the particle is m and its velocity is v, it will have an associated de Broglie wave with the wavelength λ = h/mv. Of particular interest to us is the value of the energy of the particle. Since no forces are acting on the particle, the only energy it has is its kinetic energy, $\frac{1}{2}mv^2$.

Rearranging equation (3.1), we easily obtain an expression for the velocity:

$$v = \frac{h}{m\lambda}$$

Thus, the energy E is

$$E = \tfrac{1}{2}mv^2 = \tfrac{1}{2}m\left\{\frac{h}{m\lambda}\right\}^2$$

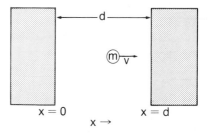

Figure 3-1. A particle in a one-dimensional box.

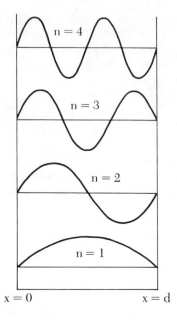

Figure 3-2. De Broglie waves associated with the particle must fit into the box exactly.

or

$$E = \frac{h^2}{2m\lambda^2} \qquad (3.2)$$

a simple relationship between the energy of the particle and its wave length.

We now assume that *the de Broglie wave associated with the particle must fit into the box exactly.* Examples of such waves are given in Figure 3-2. In each case the amplitude of the wave is zero at each wall, i.e., at x = 0 and at x = d.

Because of the restriction that the de Broglie wave must fit into the box exactly, *only certain values of the wavelength λ are possible.* Thus, for n = 1 we find that $\lambda = 2d$, since the wavelength is twice the width of the box. Similarly, for n = 2 the wavelength is equal to the width of the box, and $\lambda = d$. For n = 3, we have

$$\lambda = \frac{2d}{3}$$

and in general,

$$\lambda = \frac{2d}{n} \qquad (3.3)$$

for any value of n.

Since only those wavelengths which fit in the box are allowed, and since the wavelength in turn determines the energy, *only certain values of energy are allowed.* The quantization of energy is thus a natural consequence of the wave behavior of particles.

Substituting the value of λ from equation (3.3) into (3.2), we easily obtain

$$E = \frac{h^2}{2m\lambda^2} = \frac{h^2}{2m(n/2d)^2}$$

or

$$E = \frac{h^2 n^2}{8md^2} \tag{3.4}$$

an expression which tells us that the energy of the particle depends on the quantum number n, the size of the box, and the mass of the particle.

Figure 3-3 illustrates diagrammatically how the spacing of the energy levels depends on both the size of the box and the mass of the particle. The top section of the figure gives the lower-lying energy levels for a hydrogen atom confined to three different boxes, 10 Å, 20 Å, and 30 Å long. The bottom section gives the energy levels for a helium atom confined to the same three boxes.

The figure shows that there are two factors which make for more closely spaced levels. The first of these is the size of the box. As the box gets larger in size, the difference in energy between the levels becomes smaller. This behavior reflects the fact that the energies, as given by equation (3.4), are inversely proportional to the square of the length of the box.

The second factor affecting the spacing of the levels is the mass of the particle. For the same length of box, the energy levels are more closely spaced in the case of the heavier helium atom. Again this behavior is in accord with equation (3.4), which predicts that the energies are inversely proportional to the mass of the particle.

Although the particle in a one-dimensional box is a particularly simple example, the behavior of the energy levels in this case is in most respects typical of the behavior of the energy levels in more complicated cases. In particular, the factors influencing the *spacing* of the energy levels which we have just noted are typical of other systems. We find in general that the spacing of energy levels is determined by only *two* variables: (i) the mass of the particles involved, and (ii) what we shall call the *constraint* on them.

The first of these factors, the mass, is easily dealt with. In general, *the larger the mass* of the particle or particles, the *more closely spaced* the energy levels are. Such a result is not altogether surprising when we remember that the wave aspects of particle behavior were only noticed when scientists became interested

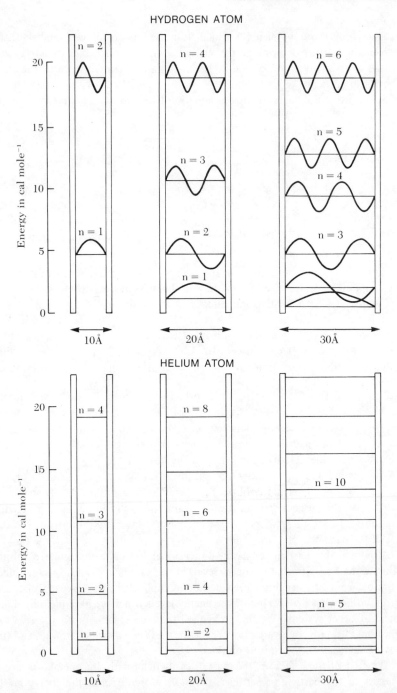

Figure 3-3. The spacing of energy levels depends on both the size of the box and the mass of the particle.

in studying particles of very small mass. Even today, the wave properties of objects with masses large enough to be weighed on a balance have never been observed, because the wavelengths are too small and the energy levels too close to each other to be detected.

The second of the two factors governing the spacing of energy levels, the *constraint*, needs some elaboration. What we mean by the constraint on a particle has two aspects to it. A particle is under a strong "constraint" if it is confined to a small region of space, or if it is acted upon by strong restraining forces. For instance, there is more constraint on the motion of an electron which is moving around a helium nucleus of charge +2 than there is on the motion of an electron moving around a hydrogen nucleus of charge +1. For any given distance from the nucleus, the electron feels twice the force of attraction from the helium nucleus as from the hydrogen nucleus. We thus expect the electron to move in a much smaller region of space in the vicinity of the helium nucleus than in the vicinity of the hydrogen nucleus; i.e., we expect the He$^+$ ion to be much smaller than the H atom. The electron, if you like, has less "elbow room" in He$^+$. The less "elbow room" a particle has, the greater the constraint on its motion.

The effect of the constraint on the spacing of energy levels is as follows: **The smaller the constraint on a particle, the more closely spaced its energy levels will be.** Again, this applies to any kind of particle in any kind of situation, and not just to those confined to a one-dimensional box. Such behavior follows naturally from the de Broglie model, and from the de Broglie relationship, equation (3.1). The smaller the constraint on a particle, the longer will be its wavelength. According to equation (3.1), the larger its wavelength, the smaller its velocity, and hence its energy, will be.

Whenever we use the de Broglie approach, we must remember that it is only an approximation to the wave nature of particles, and cannot always be applied. Sometimes, as in the case of a particle in a box, the de Broglie approach predicts the same energy levels as the more exact Schrödinger wave equation. Usually, though, it can only produce an approximate value for the energy levels. Nevertheless, as we shall soon see, such approximations are correct at least as to order of magnitude. Even if we cannot rely on the de Broglie approach to furnish *accurate* values for energy levels, we can expect it to give a good *qualitative* description of the behavior of energy levels. In particular, our generalization that a large mass and a small constraint produce more closely spaced energy levels remains valid.

TRANSLATIONAL ENERGY LEVELS

Now that we have some idea of what determines the spacing of energy levels, let us turn to some specific instances of molecular energy levels. In Chapter 2 we enumerated the four kinds of energies which gaseous molecules could possess:

translational, rotational, vibrational, and electronic. We shall now deal with each of these in turn, paying particular attention to the spacing between levels.

The translational motion of atoms or molecules refers to their movement from one part of a container to another. More exactly, it refers to the motion of their centers of gravity through space. Translational motion is least evident in solids, and is most prominent in the case of gases, where the molecules move around almost independently of each other. The translational motion of molecules in gases is constrained only by the walls of the container. Accordingly, we can treat the translational motion of a single molecule of gas as that of a particle in a box, the box being the container. Such a box will be three-dimensional, of course, rather than one-dimensional as in the case treated previously.

To take a specific instance of translational energy levels, let us consider the energy levels of an oxygen molecule in a container of volume one liter. For the sake of simplicity we shall imagine the container to be a cube with sides 10 cm long oriented in the x, y, and z directions. We shall assume that the quantization of energy in any one of these directions is independent of that in any other direction. We need then only consider one of these directions, the x-direction. Applying the results of our one-dimensional approach, equation (3.4), directly, we have

$$E_x = n_x^2 \, \frac{h^2}{8md^2} \tag{3.5}$$

in which both E and n are subscripted with an x to remind us that we are considering only one out of three dimensions.

Using cgs units, we now insert the following data into equation (3.5):

$$h = 6.63 \times 10^{-27} \text{ erg sec}$$

$$m = \frac{32}{6.023 \times 10^{23}} \text{ g}$$

$$d = 10 \text{ cm}$$

with the following result:

$$E_x = \frac{n_x^2}{8} \cdot \frac{(6.63 \times 10^{-27})^2}{10^2} \cdot \frac{6.023 \times 10^{23}}{32} \cdot \frac{\text{erg}^2 \sec^2}{\text{g cm}^2} \tag{3.6}$$

$$= n_x^2 \times 1.03 \times 10^{-33} \text{ ergs.}$$

(Remember that 1 erg = 1 g cm^2/sec^2.)

Equation (3.6) refers to the number of ergs *per molecule*. Chemically speaking, it is more useful to think in terms of energies *per mole*. Using the conversion factor

$$1 \text{ erg molecule}^{-1} = 1.44 \times 10^{16} \text{ cal mole}^{-1}$$

we have the final result:

$$E_x = n_x^2 \times 1.03 \times 10^{-33} \text{ erg molecule}^{-1} \times \frac{1.44 \times 10^{16} \text{ cal mole}^{-1}}{1 \text{ erg molecule}^{-1}}$$

$$= n_x^2 \times 1.49 \times 10^{-17} \text{ cal mole}^{-1}$$

(3.7)

Equation (3.7) enables us to calculate the spacing between the two lowest translational levels. The lowest level (with $n_x = 1$) has an energy of 1.49×10^{-17} cal mole^{-1}, while the next lowest level (with $n_x = 2$) has an energy four times as large. This means that the energy difference between these levels is $3 \times 1.49 \times 10^{-17} = 4.47 \times 10^{-17}$ cal mole^{-1}. Such a spacing is in sharp contrast to, say, the vibrational levels of H_2 shown in Figure 2-22. There, the two lowest vibrational levels are 11,790 cal mole^{-1} apart, some 10^{20} times more widely spaced than the translational levels just calculated!

Since translational energy levels are very closely spaced, we often regard the energy of translational motion as being continuous rather than quantized. The kinetic theory of gases, for instance, is derived by assuming that molecules move around in much the same way as golf balls or tennis balls. The results of this theory are fortunately not invalidated by the fact that translational motion is quantized.

The close spacing of the translational energy levels also means that a large number of levels are occupied even at room temperature. Again, this is in sharp contrast to the behavior of vibrational levels. As can be seen from Figure 2-22, even at $1000°$ K only the nine lowest vibrational levels of H_2 are occupied.

Some idea of how many translational levels are occupied at room temperature can be obtained by calculating the translational quantum number in the x-direction for an O_2 molecule of average energy at $25°$ C. The kinetic theory of gases tells us that the total translational kinetic energy of a mole of gas is $\frac{3}{2}RT$. This means that the total kinetic energy in the x-direction will be $\frac{1}{2}RT$. At $25°$C, this has the value $\frac{1}{2} \times 1.987 \times 298 = 296$ cal mole^{-1}. Inserting this quantity into equation (3.7), we obtain:

$$296 = n_x^2 \times 1.49 \times 10^{-17}$$

$$n_x^2 = 19.9 \times 10^{18}$$

(3.8)

$$n_x = 4.45 \times 10^9$$

This result means that an average molecule of O_2 at $25°$ C in a medium-sized container occupies a translational level with a very high value of n_x, in the vicinity of 10^9 or 10^{10}. The number of energy levels corresponding to the x-direction that are actually occupied under these conditions must thus be larger than 10^9 or 10^{10}. Similar remarks apply to the y and z directions.

Since the number of energy levels which are occupied in a given set depends on the spacing between the levels, it is convenient to have some means of comparing these spacings. This is usually done by comparing the distance between two levels with the quantity RT. The vibrational levels of H_2 are spaced some 11,790 cal mole^{-1} apart. At 298° K, the quantity RT is equal to $1.987 \times 298 = 592$ cal mole^{-1}. At this temperature we could then say that the spacing was equal to $(11,790/592)$ RT = 19.9 RT. A similar calculation for the translational levels of O_2 considered above gives a value of 7.5×10^{-20} RT for the spacing of the two lowest levels at the same temperature.

Table 3-1

The variation of number of occupied levels with the spacing of levels.
(Figures are for equally spaced levels and a mole of particles.)

Spacing of levels in terms of RT	Number of levels with a population larger than one particle
100 RT	only ground state
10 RT	6
RT	55
10^{-1} RT	525
10^{-2} RT	5100
.	.
10^{-20} RT	10^{21}

Table 3-1 gives us an idea of how the number of levels which are occupied varies with the spacing of levels. The numbers apply to the distribution of one mole of particles among equally spaced energy levels. A level is considered to be occupied if its population (as calculated from the Boltzmann law) is more than one, and unoccupied if the calculation indicates that there is less than one particle in it.

As can be seen from the table, if the spacing is greater than 100 RT, only the ground state will be occupied. If the spacing is in the region of 10 RT to 10^{-1} RT, less than a thousand levels are occupied. If the spacing is very much less than 10^{-1} RT we can expect a very large number of levels to be occupied.

In general, energy levels are not evenly spaced as assumed in Table 3-1. Accordingly, we can expect the table to give only a rough (order of magnitude) idea of the number of occupied levels. Certainly, it leads us to expect that a very large number of translational levels will be occupied at normal temperatures and pressures.

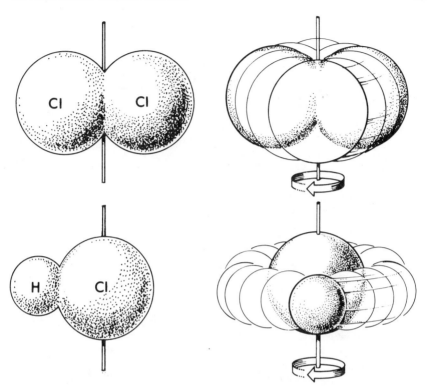

Figure 3-4. Chlorine molecule and hydrogen chloride molecule rotating about their centers of gravity.

ROTATIONAL ENERGY LEVELS

Molecules are free to rotate about their centers of gravity. Two examples of rotation are shown in Figure 3-4, for Cl_2 and for HCl. When Cl_2 rotates, both nuclei move to the same extent, since the center of gravity is equidistant from the two identical nuclei. In HCl, by contrast, the center of gravity is very close to the Cl nucleus since the Cl nucleus is some 35 times heavier than the H nucleus. To a first approximation, we can regard the rotation of an HCl molecule as the rotation of an H atom around the Cl atom.

With this approximation, let us now calculate the rotational levels of HCl using the de Broglie approach. Here there is no box into which to fit the hydrogen atom "wave." Instead, we must fit such a wave into the circumference of the circle which the H nucleus describes as it moves around the Cl nucleus. Effectively, this means that we take the circumference, $2\pi r$, as being equal to an integral number of wavelengths; i.e., we assume that

$$n\lambda = 2\pi r$$

or

$$\lambda = \frac{2\pi r}{n}$$

Substituting this value of λ into equation (3.2), we then obtain:

$$E_n = \frac{h^2}{2m\lambda^2}$$

$$= \frac{h^2}{2m\left(\frac{2\pi r}{n}\right)^2}$$

or

$$E_n = \frac{n^2 h^2}{8\pi^2 m r^2} \qquad \text{(3.9)}$$

Substituting r = 1.275 Å, the experimentally determined HCl internuclear distance, and taking the mass of hydrogen as 1 amu, we have:

$$E_n = \frac{n^2 (6.63 \times 10^{-27})^2}{8\pi^2 (1.275 \times 10^{-8})^2} \cdot \frac{6.023 \times 10^{23}}{1} \text{ erg molecule}^{-1}$$

$$= 2.06 \times 10^{-15} n^2 \text{ erg molecule}^{-1}$$

$$= (2.06 \times 10^{-15})(1.44 \times 10^{16}) n^2 \text{ cal mole}^{-1}$$

$$= 29.7 n^2 \text{ cal mole}^{-1}$$

Thus, using the de Broglie approach, the energy of the ground state is 29.7 cal mole^{-1}, and that of the first excited state four times this value. The energy difference between the ground state and the first excited rotational state is predicted to be 3 X 29.7 = 89.1 cal mole^{-1}.

The energy difference between these two levels can be determined experimentally from the infrared spectrum of HCl. The experimental value is 60.0 cal mole^{-1}, reasonably close to the value just calculated. The exact result can be obtained by solving the Schrödinger equation, which yields the formula*

$$E_n = n(n+1) \frac{h^2}{8\pi^2 M r^2} \qquad \text{(3.10)}$$

---

*In equation (3.10) n can have the value of zero as well as 1, 2, 3, . . . etc.

in place of equation (3.9) for the rotational levels. The quantity M in equation (3.10) is the so-called *reduced mass,* and is given by

$$M = \frac{m_1 m_2}{m_1 + m_2}$$

where m_1 is the mass of the H atom and m_2 that of the Cl atom. The reduced mass makes allowance for the fact that the two nuclei are both moving around a common center of gravity. The quantity r in equation (3.10) is again the internuclear distance.

In order to familiarize ourselves further with the spacing of rotational levels, let us now consider Figure 3-5, where the experimentally determined rotational levels of five diatomic molecules are given.

We notice that the relative spacing of these levels is exactly what we would predict from the rules given above about the effect of mass and constraint on the spacing. The masses of the molecules decrease in the order I_2, Br_2, Cl_2, HCl, H_2, and this is also the order of decreasing internuclear distance. We thus expect that the spacing of the levels will get wider as we go from I_2 to H_2, in the direction of lighter and smaller molecules. Particularly noticeable is the very much wider spacing of the levels for the two hydrogen-containing species. This is a result of the small mass of the hydrogen nucleus.

The spacing of rotational levels is sufficiently close for a moderate number of them to be occupied at 25° C. The two lowest levels of Cl_2, for instance, are 1.38 cal mole^{-1} apart, or 2.3×10^{-3} RT apart at this temperature. Reference to Table 3-1 suggests that more than 5,000 levels would be occupied. The correct figure is closer to 200. The discrepancy is due in part to the fact that Table 3-1 is based on the assumption of equally spaced levels. As can be seen from Figure 3-5, though, rotational levels are *not* evenly spaced. The lower levels are much closer to each other than the higher ones.* This is, of course, a consequence of the fact that E is proportional to n(n+1); as n increases; the difference in energy between successive levels must also increase.

VIBRATIONAL ENERGY LEVELS

Before we can discuss the vibrational energies of molecules we must first have a clear idea of the forces which operate as vibration occurs. Figure 3-6 illustrates a covalent bond in a diatomic molecule. The two nuclei are shown, together with the probability cloud corresponding to the bonding pair of

---

*The occupation of rotational energy levels by molecules has further complications which we cannot pursue in this book. One complication is that all rotational levels, apart from the ground level, are degenerate. Another complication is that the Pauli principle prohibits certain levels from being occupied if both nuclei are identical, and if the spins of the nuclei are not correctly oriented with respect to each other.

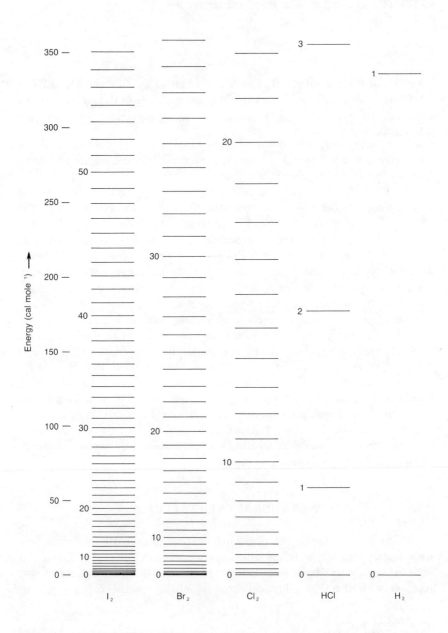

Figure 3-5. Rotational energy levels of five diatomic molecules; values were experimentally determined.

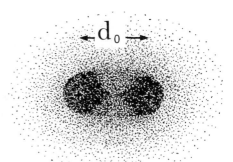

Figure 3-6. In a covalent bond in a diatomic molecule, the probability cloud is densest between the nuclei. The equilibrium internuclear distance is d_0.

electrons. This electron probability cloud is, on the whole, more dense *between* the nuclei than elsewhere. There is thus a concentration of negative charge between the positive nuclei. The two nuclei are held together despite the force of their mutual repulsion by the attraction of this intervening cloud.

At the so called equilibrium distance, d_0, between the two nuclei, these forces of attraction and repulsion balance exactly. If the molecule is now "squeezed" so that the internuclear distance d becomes less than d_0, this balance of attractive and repulsive forces is upset. As d decreases beyond d_0, the forces of repulsion between the nuclei become much larger than the attractive forces. It thus requires energy to decrease the internuclear distance below d_0. In other words, the more we squeeze the molecule, the higher its potential energy becomes.

We can also argue that the more we *stretch* the molecule, the higher its potential energy will be. At internuclear distances larger than d_0, the internuclear repulsion becomes quite small. As a result, the attractive forces between nuclei and cloud predominate. In order to stretch the molecule, we need to expend energy to overcome these attractive forces. If enough energy is exerted, it is possible to separate the two nuclei entirely, and two separate atoms will result. Thus, although the potential energy of the molecule increases with stretching, it will increase only up to a certain value, the so-called dissociation energy. Beyond this value the molecule will dissociate. In Figure 3-7, we show how the potential energy of the diatomic molecule HCl varies with the internuclear distance d.

When a molecule vibrates, it is subjected to the balance of forces just described. Let us consider such a vibration as beginning when the molecule is at "full stretch," i.e., when the internuclear distance is at its greatest just before the atoms start moving together again. Such a situation would be given by the point A in Figure 3-7 if the molecule had a vibrational energy of magnitude E_V.

At the exact moment of full stretch, the nuclei will be stationary, but will be attracting each other. Under the influence of this attraction they will accelerate toward each other, gaining kinetic energy and losing potential energy until the

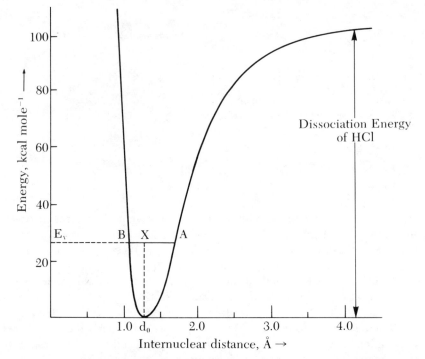

Figure 3-7. Potential energy of an HCl molecule as a function of internuclear distance. The equilibrium distance is at X, the minimum of the curve.

distance between them is d_0 at the point X. At this point the attractive force will be exceeded by the repulsive force and the atoms will begin to slow down. Eventually the repulsive forces will bring the atoms to a complete halt. The molecule will then be at "full squeeze" at point B. The repulsive force will continue to operate and will now push the two atoms apart, until the distance d_0 is again reached at X. After this, the attractive force will take over once more, slowing the atoms down till "full stretch" is again achieved. The whole process will then repeat itself indefinitely until interfered with, perhaps by a collision from another molecule, which may change the energy of the molecule.

The picture of vibration we have just given is, of course, a classical rather than a quantum description, since the wave aspects of the particles are not featured. It would be nice if we could now introduce the de Broglie relationship as we did for a particle in a box. If we try to do this, though, we immediately encounter a difficulty. The de Broglie relation we wish to introduce into our calculation,

$$\lambda = \frac{h}{mv}$$

relates the "wavelength" of the vibration to a velocity. Since the velocities of the particles are constantly changing as the molecule vibrates, it is difficult to know which velocity we should use. We could perhaps resort to using an average velocity, but the truth is that our difficulty points up a real deficiency in the de Broglie approach. It cannot handle situations in which the velocity varies. Attempts to broaden the scope of de Broglie waves in this direction lead to the Schrödinger wave equation, and, alas, to a lot of mathematics as well.

Despite such difficulties, we can still use the de Broglie approach to get an *approximate* idea of the spacing of the vibrational energy levels by using the "particle in a box" model as a rough description of vibrational motion. As an example let us take the case of HCl. Here we can regard the heavy Cl atom as coinciding with the center of gravity and hence remaining stationary, while the light H atom moves toward it and away from it. We assume that this H atom behaves like a particle in a box; i.e., its velocity and kinetic energy, instead of continually fluctuating from zero to a maximum value, remain constant as the H atom moves in and out.* The size of the box can be taken as 0.5Å, since it appears from Figure 3-7 that this is about the amplitude of vibration the molecule will have if its vibrational energy is low.

Inserting the mass of the H atom, and the above value for the size of the box, into equation (3.4) we find that:

$$E_n = \frac{n^2 h^2}{8md^2}$$

$$= \frac{n^2}{8} \cdot \frac{(6.63 \times 10^{-27})^2}{(0.5 \times 10^{-8})^2} \cdot \frac{6.023 \times 10^{23}}{1} \text{ ergs molecule}^{-1}$$

$$= 1.32 \times 10^{-13} \, n^2 \text{ ergs molecule}^{-1}$$

$$= 1.32 \times 10^{-13} \times 1.44 \times 10^{16} \text{ cal mole}^{-1}$$

$$= 1.91 \, n^2 \text{ kcal mole}^{-1}$$

Our prediction is that the energy of the ground vibrational state is 1.91 kcal mole^{-1}, and that of the first excited state is four times this quantity. The spacing between these two levels should thus be $3 \times 1.91 = 5.73$ kcal mole^{-1}. The experimental value, obtained from spectra, is actually 8.3 kcal mole^{-1}. Although our result is some 40 per cent in error, it is of the right order of magnitude, which is satisfactory enough for our purposes.

In Figure 3-8, the spectroscopically determined vibrational levels for five diatomic molecules are given. These are the same five molecules whose rotational levels are shown in Figure 3-5. Perhaps the first thing we should notice about these vibrational levels is how much more widely spaced they are than the

---

*If you like, the H atom is "rattling" in its box.

equivalent rotational levels, roughly by a factor of one hundred. (The energy scales in the two figures are different: one is 100 times as large as the other.) This difference in spacing reflects a difference in constraint in the two cases, rather than a difference in mass. For example, in HCl the H atom is confined to about 0.4 Å while vibrating, but to $2\pi \times 1.275 = 8$ Å when rotating.

The spacing of the vibrational levels shown in Figure 3-8 varies from molecule to molecule in a way which is not difficult to explain. Let us first compare the vibrational levels of I_2 and Br_2. The I_2 levels are more closely spaced. This is primarily a mass effect. Heavier atoms are vibrating in I_2 than in Br_2. There is also a difference in the constraint on the vibration in the two cases. The forces acting on the I atoms are smaller than those acting on Br atoms. In I_2 the nuclei are farther apart than in Br_2 and the bonding electron pair cloud more diffuse. Thus, both attractive and repulsive forces are smaller in the larger molecule. In a similar manner, one can explain the fact that the Cl_2 levels are more closely spaced than the Br_2 levels.

The two remaining molecules in Figure 3-8 are HCl and H_2. The spacing of the vibrational levels in these two molecules is significantly wider than for the halogens. This is largely due to the very light H atom involved in the vibration. The forces constraining the vibration are also larger than for the halogens; as we shall see in Chapter 5, both the H–Cl and H–H bonds are stronger than any halogen-halogen bond.

A minor feature of Figure 3-8, which is perhaps worth noticing, is that the vibrational levels are not quite evenly spaced. A close inspection of the figure reveals that the higher levels are a little closer to each other than the lower levels. When molecules are in the higher vibrational states, the two atoms spend a large portion of their time at some distance from each other. At such large distances the force of attraction between atoms is smaller, and the constraint is less.

Since vibrational levels are more widely spaced than rotational levels, we can expect fewer vibrational levels than rotational levels to be occupied at the same temperature. The spacing between the two lowest vibrational levels of I_2, for example, is 624 cal mole^{-1} or 1.05 RT at 25° C. Reference to Table 3-1 suggests that a little more than 55 of these levels will be occupied. By contrast, only the three lowest of the more widely spaced H_2 vibrational levels are occupied at 25°C.

ELECTRONIC ENERGY LEVELS

The fourth and last type of energy in which we are interested is that associated with the movement of electrons around a nucleus or several nuclei, in atoms, molecules, and ions. We shall deal only very briefly with this kind of energy for two reasons. In the first place, you are already familiar with many aspects of this subject from a study of the electronic structure of atoms and of chemical bonding. In the second place, since we are interested primarily in distributions of particles over energy levels, we need only establish that the

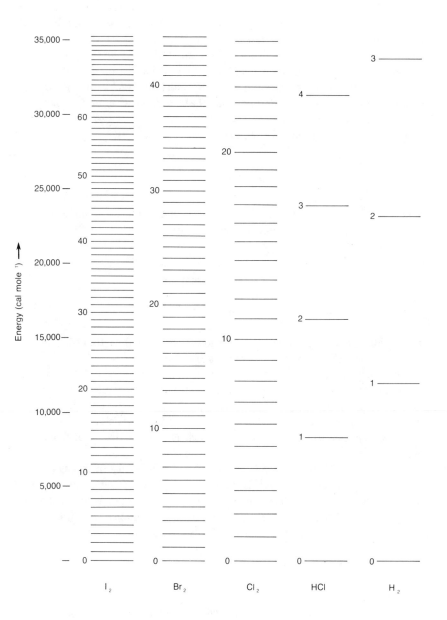

Figure 3-8. Vibrational energy levels of five diatomic molecules; values were experimentally determined.

spacing of electronic levels is so wide that only the ground state is occupied. In general, excited states are too high in energy to be occupied, except at quite high temperatures.

To obtain a rough idea of the spacing of electronic energy levels, let us use the de Broglie approach to calculate the spacing between the two lowest levels of the valence electron in the potassium atom. We can regard this electron as moving in a circular orbit around the nucleus and deal with it in the same way as we dealt with the rotation of an H atom around a Cl atom, on page 59. Just as we assumed there that the H atom's "wave" must fit exactly on the circumference of its orbit, we assume here that the electron's wave must fit exactly onto the circumference of its orbit.

With such an assumption, we can again use equation (3.9):

$$E_n = \frac{n^2 h^2}{8\pi^2 \, mr^2}$$

We now insert the experimentally determined radius of the potassium atom, 2.3 Å (2.3×10^{-8} cm), and the mass of the electron, $m = 9.1 \times 10^{-28}$ g. We then obtain

$$E_n = 1.15 \times 10^{-12} \, n^2 \text{ erg molecule}^{-1}$$

$$= 16.6 \, n^2 \text{ kcal mole}^{-1}$$

The energy difference between the two lowest levels is thus $16.6 \, (2^2 - 1^2) = 49.8$ kcal mole^{-1}. The experimental value is 36.1 kcal mole^{-1}.

Electronic energy levels are usually even more widely spaced than in the potassium atom; 100 kcal mole^{-1} is a representative value. Electronic energy levels are thus appreciably more widely spaced than vibrational levels. This difference in spacing reflects the different masses of the particles involved in the two cases (an electron as opposed to a nucleus), rather than a difference in constraint.

The fact that electronic levels are usually so widely spaced means that, in general, only the ground state is occupied at ordinary temperatures. The representative value of 100 kcal mole^{-1} given above corresponds to a spacing of 169RT at 25°C. Such a wide spacing is more than large enough to insure that no excited levels are occupied, as a glance at Table 3-1 shows.

SUMMARY

In this chapter we have been concerned with the *spacing* of energy levels. We have seen that there are two factors which determine this spacing. First, there is the mass. The heavier a particle is, the *more closely spaced* will be the energy levels which it can occupy. Second, there is the constraint. The *less* a particle is constrained, the *more closely spaced* will be the energy levels which it can occupy.

The energy levels which a molecule in a gas can occupy are of four kinds: translational, rotational, vibrational, and electronic. Their spacing increases in the order given: the most closely spaced levels are translational levels, while the most widely spaced levels are electronic levels. Translational levels are some 10^{-17} cal mole^{-1} apart, while electronic levels are 10^5 cal mole^{-1} apart. Such a large difference in the spacing of levels is explained by the fact that translational levels correspond to the motion of heavy molecules confined to a rather large volume, while electronic levels correspond to the motion of very light electrons confined to the tiny space occupied by a molecule or an atom.

We are interested in the spacing of energy levels because it is one of two factors which determine the number of levels which are occupied by atoms or molecules. The other factor influencing this quantity is, of course, the temperature. We find that for a given temperature, the number of levels which are occupied increases as the levels get more and more closely spaced.

The most convenient way of relating the spacing between levels to the number of levels occupied is to compare the spacing with the quantity RT. If the spacing is of the order of 100 RT or larger, only the ground state will be occupied. If the spacing lies between 10 RT and 10^{-1} RT, less than a thousand levels will be occupied. If the levels are more closely spaced than 10^{-1} RT, we can expect a very large number of levels, much greater than a thousand, to be occupied.

Using this comparison with RT as a criterion, we find that at 25°C, the usual reference temperature, the different kinds of molecular energy levels are occupied to very different extents. As expected, we find that the largest number of occupied levels occurs in the case of those that are most closely spaced, the translational levels. Typically, a very large number of these levels are occupied, as many as 10^{20} or more. Not nearly so many rotational and vibrational levels are occupied at 25°C. The number varies from a few thousand in the case of the rotation of a heavy molecule to only three in the case of the vibrational levels of H_2.

The smallest number of occupied levels occurs with electronic levels. Typically these are so far apart that only the ground state is occupied. 25°C is too low a temperature for even one or two molecules per mole to be electronically excited.

PROBLEMS 3

1. Calculate the de Broglie wavelength of
   (a) a man of mass 100 kg walking at a velocity of 5 km per hour.
   (b) a nitrogen molecule moving at 500 m sec^{-1}.
   (c) an electron moving at velocity of 2×10^8 cm sec^{-1}.

2. Electron microscopes utilize the wave properties of moving electrons. In order to be of use in this connection, an electron must have a wavelength which is less than about 0.1 Å. How fast must the electron be traveling in order to have such a small wavelength?

3. Calculate the energies (in cal mole^{-1}) of the six lowest levels of a helium atom confined to a one-dimensional box 30 Å long.

4. Calculate the energies (in cal mole^{-1}) of the six lowest levels of a helium atom confined to a one-dimensional box 20 Å long.

5. Make a rough estimate of how widely spaced the energy levels are in nuclei by using a "particle in a box" approximation. Calculate the two lowest levels when a proton is confined to a one-dimensional box 10^{-12} cm long. Express the answer in kcal mole^{-1}.

6. Arrange the following gases in order of increasingly widely spaced translational levels. (Assume the same volume in each case.)
   (a) Xe, Ne, He, Kr.
   (b) CH_4, F_2, SF_6, HI.

7. The two lowest rotational levels for HCl are 60 cal mole^{-1} apart. Assuming (incorrectly) that these levels are evenly spaced, estimate from Table 3-1 how many of these levels will be occupied at 25°C.

8. Calculate the difference in energy between the ground state and the first excited rotational state of HBr using equation (3.9). The H-Br distance in this molecule is 1.41 Å. Compare this value with the observed value of 47.8 cal mole^{-1}, and with the value obtained from equation (3.10).

9. Calculate the energy difference between the ground state and the first excited rotational state of I_2 using equation (3.10). The I-I distance in I_2 is 2.67 Å. Compare this value with that obtained for H_2. The H-H distance in H_2 is 0.742 Å.

10. Arrange the following molecules in order of increasingly widely spaced rotational levels.
   (a) F_2, N_2, O_2, Cl_2.
   (b) CF_4, CH_4, CBr_4, CI_4.
   (c) H_2O, H_2S, H_2, SF_6.

11. The spacing between the vibrational levels of the hydrogen halide molecules decreases from HF (11.3 kcal mole^{-1}), through HCl and HBr down to HI (6.4 kcal mole^{-1}). Account for this observation qualitatively.

12. The difference in energy between the ground state and the first excited vibrational state of the HCl molecule is 8.25 kcal mole^{-1}. Using a "particle in a box" approximation, calculate the equivalent difference for DCl. Assume that the constraint on the H and the D atoms is identical in the two cases. Compare your result with the

observed value of 5.98 kcal mole^{-1}.

13. Suggest why it requires less energy to separate the two atoms in F_2 completely than to separate the two atoms in N_2. Which molecule would you expect to have the more closely spaced vibrational levels, F_2 or N_2?

14. The vibrational levels of HI are approximately evenly spaced at an interval of 6.4 kcal mole^{-1}. From Table 3-1, estimate how many of these levels will be occupied at 25°C.

15. The radius of a sodium atom is about 1.9 Å. Use this quantity to calculate the energy required to excite the valence electron in the Na atom from the ground state to the first excited state. Compare the answer obtained to the experimental value of 48.3 kcal mole^{-1}. (This transition, in reverse, gives rise to the well-known yellow line in the spectrum of Na.)

Chapter 4

THE COMPETITION BETWEEN ALTERNATIVE SETS OF ENERGY LEVELS

The last two chapters have been concerned almost exclusively with energy levels. In Chapter 2 we dealt with the Boltzmann distribution law, which governs the occupation of energy levels, and in Chapter 3 we considered those factors governing the spacing of energy levels. Such a background enables us to return to the subject of chemical equilibrium and to look at it in terms of energy levels. For the sake of simplicity we shall confine ourselves to equilibria in which all the reactants and products are *gases*.

ISOMERISM

We shall start by interpreting what is probably the simplest of all chemical equilibria, the equilibrium between *optical isomers,* in terms of energy levels and distributions. Before we can do this we must first learn what is meant by the terms isomerism and optical isomerism.

In an equilibrium of the type

$$A(g) \rightleftharpoons B(g) \qquad\qquad (4.1)$$

in which only two chemical species, A and B, are involved, the two species must have not only the same molecular weight but also the same molecular formula. Species related in this way are said to be *isomers.*

As an example of isomerism we take ethyl alcohol and dimethyl ether. A molecule of ethyl alcohol has the following arrangement of bonds between its atoms:

$$
\begin{array}{cc}
H & H \\
| & | \\
H-C-C-O-H \\
| & | \\
H & H
\end{array}
$$

A molecule of dimethyl ether has a different arrangement, namely the following:

$$
\begin{array}{cc}
H & H \\
| & | \\
H-C-O-C-H \\
| & | \\
H & H
\end{array}
$$

Despite these different arrangements, both ethyl alcohol and dimethyl ether have the same molecular formula, C_2H_6O, and are therefore isomers.

Differences between isomers need not be as obvious as the example just quoted. Isomers can have the same atoms linked in the same sequence but yet be different *geometrically.* Sometimes this difference can be very subtle. On the whole, the more subtle the difference between two isomers, the simpler it is to explain an equilibrium between them. The simplest examples of chemical equilibrium must necessarily occur between almost identical molecules. If two species which differ only slightly from each other are in equilibrium with each other, it is the *differences* between them rather than their similarities which determine the equilibrium position.

OPTICAL ISOMERISM

Our first example of chemical equilibrium is that between so-called *optical isomers.* Figure 4-1 illustrates the two optical isomers of bromochlorofluoromethane, using "ball and stick" models.

At first sight these two molecules, which have been labeled A and B for easy reference, might appear identical. In fact they are not. A and B, which are said to have different *configurations,* are related in the same way as a left-hand glove is to a right-hand glove. Not only are they mirror images of each other, but no amount of rotation about any axis will convert the one form into the other, just as no amount of rotation of any kind will convert a left-hand glove into a right-hand glove. A can be converted to B only by some process such as the breaking of a bond or a drastic distortion or flattening of the molecule. Without doubt, the best way of appreciating this point is for you to construct models of both isomers for yourself, using one of the molecular model kits that are

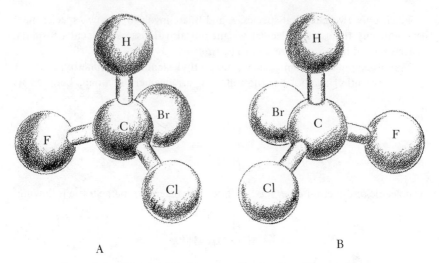

A   B

Figure 4-1.   The two optical isomers of bromochlorofluoromethane.

available. You will then see that the two models may be interconverted only by partially dismembering them.

The "left hand versus right hand" difference between the molecules A and B is such a subtle one that almost all the properties of the two isomers are identical. One property in which the two isomers differ is their response to polarized light: hence the term *optical* isomerism. The details of such behavior need not concern us here. We need only note that it is possible, by using polarized light, to analyze how much of each optical isomer is present in a mixture of the two.

Of greatest interest to us in this discussion are the energy levels which the molecules of the two isomers can occupy. We can easily argue that these will be identical in all respects. If both isomers occupy the same container, the translational constraint on their molecules will be the same in both cases. The *masses* of A and B molecules are also the same. Thus the translational energy levels of the two isomers must be identical. The A and B molecules will also have identical rotational levels, since all interatomic distances and all angles are identical in molecules which are mirror images of each other. Similarly, the forces of attraction between corresponding pairs of atoms in the two molecules will also be the same. Hence, we conclude that the vibrational levels of A and B will also be identical.

Consider now the equilibrium between the two isomers of gaseous bromochlorofluoromethane, at some temperature T above its boiling point, and at one atmosphere pressure:

$$A(g) \rightleftharpoons B(g) \quad (T°C, 1 \text{ atm.}) \tag{4.2}$$

For the sake of the argument, let us assume that A and B may be readily interconverted and equilibrium rapidly established. In actual fact that is not possible in the gaseous state, although suitable catalysts are available if the equilibrium is studied in solution.

In such an equilibrium situation, not only are molecules constantly exchanging energies, but they are also constantly interconverting from configuration A to configuration B and vice versa. We might find, for instance, that at a particular instant a given molecule has the A configuration, a small rotational energy, and a large translational energy. A little later the same molecule could well have the B configuration, a large rotational energy, and only a moderate translational energy. In applying the Boltzmann law to such a situation, we must have regard for both the A set of energy levels and the B set. In this way we take into account every possible combination of energy and configuration a molecule might have.

Figure 4-2 represents such a Boltzmann distribution of molecules over the two sets of levels. This representation is schematic only: the levels shown are not intended as an accurate representation of the translational, rotational, or vibrational levels, or any combination of them. The figure does, however, emphasize that for each and every A level there is a corresponding B level of equal energy. The thickness of the lines representing the levels in the figure is proportional to the population of that level, the lower levels being more heavily populated than those of higher energy.

Notice that for every A level there is. a corresponding B level of *equal population.* This is a direct consequence of the Boltzmann law and the fact that for every A level there is a B level of equal energy. Summing populations of the levels of each isomer, it is clear that the total number of molecules with the A configuration must be equal to the total number of molecules with the B configuration. The equilibrium state thus corresponds to equal concentrations of both A and B species. Accordingly, the equilibrium constant K_c is unity for reaction (4.2).

No example of *gaseous* equilibrium between optical isomers appears to have

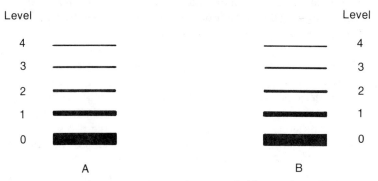

Figure 4-2. Boltzmann distribution for two optical isomers in equilibrium.

been studied experimentally. Nevertheless, a large number of such equilibria have been studied in *solution*. Invariably, the equilibrium state has been found to consist of *equal concentrations* of both isomers *at all temperatures* unless the solvent itself is an optical isomer. Such a result gives us confidence in the argument we have used, even though an equilibrium in solution is not quite the same as the corresponding gaseous equilibrium and would have to be treated somewhat differently.

THE COMPETITION BETWEEN ALTERNATIVE SETS OF LEVELS

The approach we have just used in dealing with optical isomerism, in which we applied the Boltzmann law to two sets of energy levels simultaneously, is a useful way of looking at all chemical equilibria involving gases. We can regard such a chemical equilibrium as a *competition between two alternative sets of energy levels*. One set of levels corresponds to the reactants, and another to the products. The problem of deciding whether reactants or products will be favored at equilibrium then becomes one of deciding which of the alternative sets of levels will be more populated.

In the equilibrium between optical isomers just discussed, both sets of levels involved in the competition for population were identical. Since there was nothing to choose between the two sets, we found them to be equally populated at equilibrium.

In most chemical equilibria the two competing sets of energy levels will *not* be identical. They will differ in a variety of ways. Our problem is to discover what kinds of differences between sets of energy levels will be important in their effect on chemical equilibrium, i.e., in their effect on the competition for population.

It is possible to solve this problem in a completely abstract way, by considering the general problem of applying the Boltzmann law to *any* two competing sets of levels. The complexity of the algebra required for such a treatment puts it beyond the scope of this book. Despite the complexity of the problem, however, the answer is a surprisingly simple one. We find that *only three factors* are involved in determining the outcome of a competition between two sets of energy levels. The three factors involved are the following:

(a) *The relative energies* of the two alternative sets of levels.

(b) *The relative spacing* of the two alternative sets of levels.

(c) *The temperature.*

We can understand why it is that these three factors affect the position of chemical equilibrium by considering two simple, idealized examples of competing sets of levels. The first of these examples is the two sets of levels A and B shown in the left-hand box of Figure 4-3. We will suppose that these levels correspond to two isomers A and B (*not* optical isomers) which are in equilibrium with each other. The levels of the A set are regularly spaced at intervals of

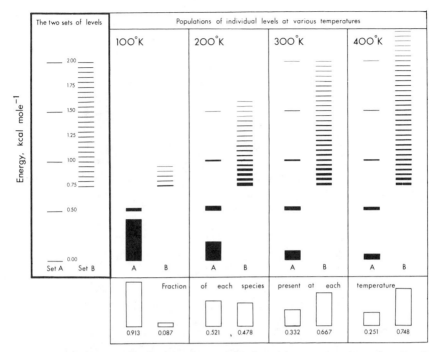

Figure 4-3. Boltzmann distribution of one mole of particles over two alternative sets of levels at various temperatures. At low temperatures set A contains most of the particles, while at high temperatures set B contains most of the particles.

500 cal mole^{-1}. The levels of set B are also regularly spaced but the spacing is one tenth that of set A, the interval being only 50 cal mole^{-1}. Another difference between the two sets of levels is the relative energies of the ground states. The ground state of A is lower in energy by 750 cal mole^{-1} than the ground state of B.

Let us now assume, as we did in the case of optical isomers, that A and B can interconvert freely. As before, we may apply the Boltzmann law to both sets of levels taken together. The results of such calculations at four different temperatures are also shown in Figure 4-3. The population of each level is represented by the thickness of the line.

At absolute zero, only one energy level, the lowest, can be occupied. Since the lowest level belongs to set A, *all* the molecules will thus be of type A at this temperature. At 100° K, as shown in the figure, the situation is not very much different from that at absolute zero. Almost all the molecules are still in the lowest level, though a few of the levels of higher energy are also occupied. Almost all of the molecules (91.3 per cent) are of type A because the two levels of lowest energy correspond to this isomer. A few of the B levels are also populated, but only sparsely since they are higher in energy. Only 8.7 per cent of the molecules are of type B at this temperature.

As the temperature rises above 100° K, more and more of the higher levels become occupied at the expense of the lower levels. Since more of these higher levels belong to B rather than A, this means that with increasing temperature, B levels are becoming more populated at the expense of A levels. The equilibrium mixture thus becomes richer in B as the temperature rises.

This process is clearly indicated in the figure. Whereas at $100°$ K only 8.7 per cent of the equilibrium mixture consists of molecules of B, at $200°$ K this percentage has risen to 47.8. At $400°$ K, B molecules comprise almost three quarters of the equilibrium mixture.

Although the percentage of B present at equilibrium rises with temperature, it can never become 100 per cent; that is, A levels can never become totally depopulated. As we have already seen in Chapter 2, when the temperature approaches infinity, the Boltzmann factor

$$e^{-\epsilon_i/kT}$$

approaches unity for all energy levels. This means that the populations of all levels approach each other at very high temperatures. Since there are ten B levels to each A level, we can expect that as the temperature approaches infinity, the ratio of the number of B molecules to A molecules will approach the limit of 10 to 1.

The composition of the equilibrium mixture will thus change markedly with temperature, in sharp contrast to the equilibrium between optical isomers. While at low temperatures species A is favored (the equilibrium composition being 100 per cent A at absolute zero) with increasing temperature B becomes more favored. At high temperatures the proportion of B increases toward the limiting value of 90.9 per cent.

It is not difficult to see why one species should be favored at low temperatures and the other at high temperatures. At low temperatures, where $e^{-\epsilon_i/kT}$ is a very tiny number for any reasonable value of ϵ_i, the lowest levels are the only ones having an appreciable population; thus the species with the *lowest energy levels* will be favored at low temperatures. As the temperature rises, the relative energies of different levels become less important in determining populations. Not only do levels of higher energy become more populated, but they become more populated at the expense of lower lying levels. As the populations of the levels begin to even out in this way, so the *relative number* of levels in the two competing sets becomes increasingly important. The species with the *more closely spaced levels* thus is favored at higher temperatures.

The arguments we have just used do not depend for their validity on the exact details of the two competing sets of levels, since they derive basically from the Boltzmann law. They apply equally well to a second example of competing sets of energy levels discussed below.

The two sets of competing levels involved in this example are shown in Figure 4-4. Both sets of levels are evenly spaced, as in our first example. The A set of levels is more closely spaced, at intervals of 50 cal mole^{-1}, while the B set

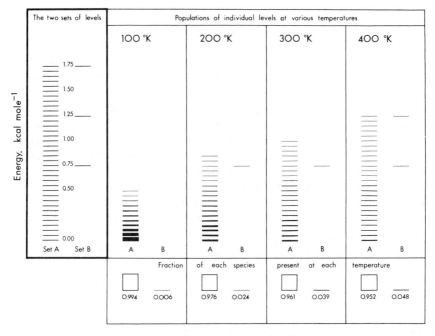

Figure 4-4. Another Boltzmann distribution of one mole of particles over two alternative sets of levels at various temperatures; this time, set A contains most of the particles at all temperatures.

is spaced at ten times this interval, i.e., 500 cal mole^{-1}. The intervals are the same as in our first example, but the ground state of the more closely spaced levels is now *lower* in energy, whereas previously it was higher. Because of this, the system behaves quite differently.

The figure shows the results of applying the Boltzmann law to the competing sets of levels at various temperatures. The results are in agreement with our generalizations. At low temperatures we find, as before, that it is the species with the *lower lying levels,* species A, which is favored. At high temperatures, it is the species with the *more closely spaced* levels which is favored. In this example, this is species A. Thus species A predominates at all temperatures, both low and high. It possesses two advantages over species B, *both* lower lying *and* more closely spaced levels.

The energy levels of species B are so few and of such high energy that they can never muster an appreciable population. Notice, however, that they do manage to attract a slightly larger population at high temperatures than at low temperatures. This occurs because of the progressive equalization of populations with increasing temperature. At very high temperatures, we can expect the ratio of A molecules to B molecules to approach the limit set by the relative spacing of the two sets of levels; namely 10 to 1. Thus the B species can never become more than one eleventh (9.1 per cent) of the total, no matter how high the temperature.

CHEMICAL EQUILIBRIUM IN GENERAL

The behavior exhibited at very high and very low temperatures by the two idealized examples we have just considered is common to *all* chemical equilibria in which all the reactants and products are gases, since it is a direct consequence of the form of the Boltzmann law. The exponential nature of the law means that only the lowest levels are appreciably populated at low temperatures in *any* situation. For a chemical equilibrium, viewed as a competition for population between alternative sets of energy levels, this can mean but one thing. That side of the equilibrium with the lower lying levels will be favored at low temperatures, for any chemical reaction we care to consider.

At high temperatures, by contrast, the Boltzmann law predicts a progressive equalization of the populations of different levels. It is precisely this equalization of populations which makes the relative spacing of the two competing sets the determining factor in fixing the position of a chemical equilibrium at high temperatures. We can thus formulate a rule which applies to all equilibria:

At low temperatures that side of a chemical equilibrium with the lower lying levels will be favored, while at high temperatures, that side with the more closely spaced energy levels will be favored.

What is meant by the two "sides" of an equilibrium to which this rule refers is best explained by means of an example. If the equilibrium is of the form

$$A(g) + B(g) \rightleftharpoons C(g) + D(g)$$

then the left-hand side of this equilibrium corresponds to the reactants A and B, while the right-hand side corresponds to the products C and D.

DISSOCIATION EQUILIBRIA

Without further ado, let us now apply these rules to a real rather than a theoretical example of chemical equilibrium. We will consider the dissociation of molecular iodine gas into its constituent atoms:

$$I_2(g) \rightleftharpoons 2 I(g) \tag{4.3}$$

It is not difficult to see in this case which of the two competing sets of energy levels lies lower in energy and which is the more closely spaced. Let us consider the relative energies first.

The conversion of molecular iodine into its atoms requires the breaking of a chemical bond and hence the expenditure of energy. This inevitably means that the ground state of the I_2 molecule (corresponding to an I_2 molecule in its lowest translational, vibrational, and rotational energy state) must be lower in energy than the ground state of two iodine atoms (corresponding to two I atoms in their lowest translational states).

The breaking of the I–I bond corresponds not only to changing from a low to a high energy situation, but also to a *relaxation of the constraint* on the movement of the iodine atoms. Instead of being forced to move around within a few Angstrom units of each other, the two iodine atoms are able to move around independently within the very much larger dimensions of the container. In view of our discussion in Chapter 3, this can only mean a shift to more closely spaced energy levels.

Application of our rule now predicts that low temperatures will favor the low energy side of (4.3), namely undissociated I_2. High temperatures, on the other hand, will favor that side corresponding to more closely spaced levels, namely iodine atoms. Experimentally, this is what is observed: molecular iodine dissociates at high temperatures.

Equilibrium (4.3) exhibits the same general behavior as the idealized example of Figure 4-3, and for the same general reasons. The position of the equilibrium swings from the side of lower energy to that of more closely spaced levels as the temperature increases. There are important quantitative differences, though, between the two cases. It requires some 35.6 kcal of energy to dissociate a mole of I_2 molecules. This is a very much larger quantity than the 0.75 kcal mole^{-1} which represents the difference in energy between the ground states of A and B in Figure 4-3. As a result, we find that the dissociation of I_2 into atoms requires a much higher temperature than the conversion of A into B. From Figure 4-3 we find that there are equal concentrations of reactant A and product B at a temperature in the vicinity of 200° K. An equivalent position is not reached in the I_2 dissociation equilibrium until about 1500° K.

Another important difference between the two cases lies in the relative spacing of the two sets of competing levels. When I_2 molecules dissociate, rotational and vibrational levels are replaced by translational levels. The latter levels are some 10^9 to 10^{11} times as closely spaced as those they replace. This is in sharp contrast to the 10 to 1 ratio of spacings of the example in Figure 4-3.

In our idealized example we could never get more than 90.9 per cent B, no matter how high the temperature; in contrast, the dissociation of iodine can go virtually to completion at a sufficiently high temperature.

This dissociation of I_2 into its atoms is, of course, typical of all dissociations. We could apply the same arguments to the dissociation of any diatomic molecule into its contituent atoms or of any polyatomic molecule into simpler fragments. All such processes correspond to the movement from a low energy to a high energy set of levels, and to a relaxation of the constraint on the motion of particles. We thus expect to find, and experience bears out, that at a sufficiently high temperature *any* molecule will decompose virtually completely. A molecule with very strong bonds will decompose only at high temperatures, while a loosely bound molecule will require much lower temperatures. In the sun's atmosphere, for example, at a temperature of some 6000° K, only a few molecules (such as CN) still survive, and these only at very low concentration.

ADVANTAGES AND LIMITATIONS

To look at chemical equilibrium as a competition between alternative sets of energy levels has both advantages and limitations. The major advantage of such a view is that it convinces us, once and for all, that chemical equilibrium and chemical reactions are not mere happenstances. There are basic *reasons* why some reactions go to completion and others do not, why some compounds are stable and others decompose, why at one temperature a reaction goes in one direction and at another temperature it goes in the opposite direction. The explanation of such behavior lies ultimately in the relative energies and the relative spacings of different sets of levels. Unfortunately, our view of chemical equilibrium as a competition between alternative sets of levels is often difficult to apply. Real chemical equilibria are always much more complex than the readily visualized examples of Figures 4-3 and 4-4. Consider, for instance, the equilibrium:

$$N_2(g) + 3\,H_2(g) \rightleftharpoons 2\,NH_3(g) \qquad (T^\circ\,C,\,1\,atm) \qquad \textbf{(4.4)}$$

In order to look at this equilibrium in terms of competing energy levels, we must consider the translational, rotational, and vibrational levels of each of the three species involved, and also their relative energies. Such a treatment is possible if enough spectral data are available, and leads to excellent agreement with experiment. Obviously, though, such an approach is quite complex, and beyond the scope of our discussion here.

Fortunately, there is an easy way out of this difficulty. It turns out that it is not really necessary to think in detailed terms about the energy levels involved in an equilibrium such as (4.4) in order to predict its behavior. It is possible to connect the relative energies and the relative spacings of the competing sets of energy levels to differences between *measurable properties* of the reactants and products. The relative energies of two competing sets of levels can be related to a difference in a property called the *internal energy,* while their relative spacing can be related to a difference in a property called the *entropy.*

In the two chapters which follow, we shall consider each of these two properties in turn. Having done so, we will be able to return to the subject of chemical equilibrium and reformulate it in terms of these two new concepts. Such a reformulation will enable us to apply the principles already developed to a large number of chemical reactions.

PROBLEMS 4

1. Use the fractions given at the bottom of Figure 4-3 to calculate K_C for the reaction

$$A(g) \rightleftharpoons B(g)$$

at $100°$ K, $200°$ K, $300°$ K, and $400°$ K. What is the limiting value for K_C at very high temperatures?

2. By contrast, consider the system in Figure 4-4. Again calculate K_C for the temperatures given in the figure. What is the limiting value of K_C at very high temperatures in this second case?

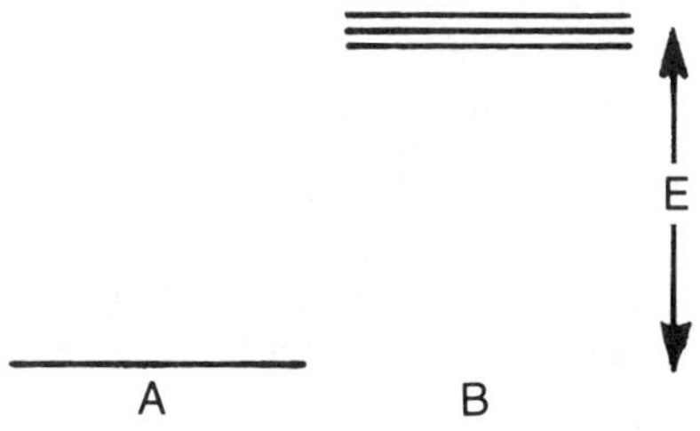

Figure 4-5. The two sets of energy levels referred to in Problem 3.

3. Consider an equilibrium between two hypothetical isomeric species A and B whose energy levels are shown in Figure 4-5. Molecule A has only one level, the ground state, while molecule B has a triply-degenerate ground state and no excited states. The energy difference between these levels is E cal mole^{-1}. Show that the equilibrium constant for the reaction

$$A(g) \rightleftharpoons B(g)$$

is given by the equation

$$K_C = 3e^{-E/RT}.$$

4. If E = 10 kcal mole^{-1} in Figure 4-5, calculate K_C at $25°$C, $500°$K, and $1000°$ K.

5. Calculate K_C for the system shown in Figure 4-5 at $25°$ C, if
   a) E = 0.1 kcal mole^{-1}
   b) E = 1 kcal mole^{-1}
   c) E = 10 kcal mole^{-1}
   d) E = 100 kcal mole^{-1}

6. Consider an equilibrium between the two hypothetical isomers C and D whose energy levels are shown in Figure 4-6. Both sets of energy

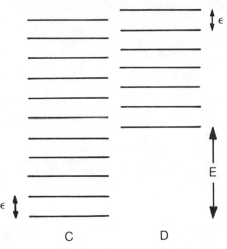

Figure 4-6. The two sets of levels referred to in Problem 6. The spacing of levels is the same for both sets.

levels are evenly spaced, the interval between levels being given by the quantity ϵ in both cases. The ground state of C is lower in energy than the ground state of D by an amount E. Show that the equilibrium constant for the reaction

$$C(g) \rightleftharpoons D(g)$$

is given by the equation

$$K_c = e^{-E/RT}$$

and is independent of the spacing ϵ, provided only that this is the same for both sets of levels.

7. In Figure 4-7, five different energy level situations are shown for an equilibrium of the type

$$A(g) \rightleftharpoons B(g).$$

In all five cases the energy levels of both species are evenly spaced. The interval between levels is either 1 cal mole^{-1} or 2 cal mole^{-1}.

Match each situation given in the figure with the five following possibilites for the behavior of K_c:

(a) K_c is zero at $0°$ K but rises to a limit of 1/2 at very high temperatures.

(b) K_c is zero at $0°$ K but rises to a limit of 2 at very high temperatures.

(c) K_c is unity at all temperatures.

(d) K_c is unity at $0°$ K and increases to a limit of 2 at very high temperatures.

(e) K_c is zero at $0°$ K but rises to a limit of unity at very high temperatures.

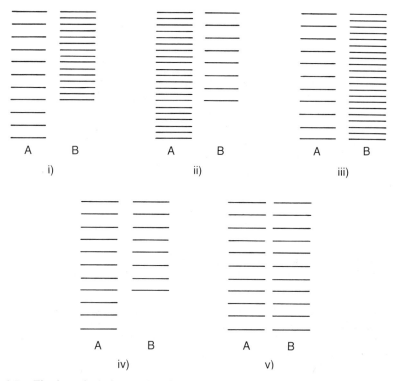

Figure 4-7. Five hypothetical examples of competing sets of energy levels. The spacing in each set is either 1 cal mole^{-1} or 2 cal mole^{-1}.

8. Under what circumstances will K_c for an equilibrium between two isomers first increase and then decrease as the temperature is raised from 0° K?

9. n-Butane and isobutane are two isomers with the formula C_4H_{10}. K_c for the equilibrium

$$n\text{-butane} \rightleftharpoons isobutane$$

has the value 4.4 at 300° K and 0.52 at 400° K. Which isomer has the lower ground state? Which has the more closely spaced energy levels?

10. Two of the isomers of xylene, C_8H_{10}, are called meta-xylene and ortho-xylene. The equilibrium constant K_c for the reaction

$$m\text{-xylene} \rightleftharpoons o\text{-xylene}$$

is found to be 0.27 at 300° K and 0.55 at 1000° K. Which isomer has the lower ground state? Which has the more closely spaced levels? (Careful!)

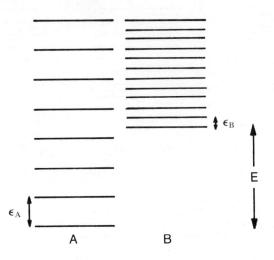

Figure 4-8.   The two sets of energy levels referred to in Problem 11.

11.   Figure 4-8 represents the energy levels of two hypothetical isomers A and B. The A set of levels are evenly spaced at an interval ϵ_A, while the B set is also evenly spaced but at an interval ϵ_B. The difference in energy between the two ground states is E.

Show that, for this system, the equilibrium constant K_c for the reaction

$$A(g) \rightleftharpoons B(g)$$

is given by the expression

$$K_c = e^{-E/RT} \left\{ \frac{Q_B}{Q_A} \right\}$$

where

$$Q_A = \frac{1}{1 - e^{-\epsilon_A/RT}}$$

and

$$Q_B = \frac{1}{1 - e^{-\epsilon_B/RT}}$$

Substituting the values ϵ_A = 500 cal mole^{-1}, ϵ_B = 50 cal mole^{-1}, and E = 750 cal mole^{-1}, cross check some of the numbers given in Figure 4-3.

Finally, show that at a very high temperature K_c approaches the limit ϵ_A/ϵ_B.

THE INTERNAL ENERGY

HEAT CHANGES

When a chemical or physical change occurs, heat energy is almost always released or absorbed. As one example among many, let us take the reaction of magnesium with oxygen:

$$Mg(s) + \tfrac{1}{2}O_2(g) \rightarrow MgO(s) \tag{5.1}$$

Many of you will be familiar with this reaction and may even have performed it in the laboratory. If the magnesium is ignited in air, it burns to the oxide with a very intense white flame. Both heat and light are evolved; the resulting oxide is at first white hot but rapidly cools to a white powder at room temperature. The amount of heat energy given off in such a reaction can be measured experimentally, as we shall see in a later chapter. It is found that the oxidation of one mole of magnesium according to equation (5.1) results in the *liberation* of 144 kilocalories of heat energy.

As a second example let us take another familiar change, the boiling of water at 100°C and one atmosphere pressure:

$$H_2O(l) \rightarrow H_2O(g) \tag{5.2}$$

As is well known, it requires heat energy from a flame, say, or an electric hot-plate, to convert liquid water to its vapor. The exact amount of heat energy needed can be measured experimentally. It turns out that 9.72 kilocalories is required to convert a mole of liquid water into a mole of water vapor at 100°C and one atmosphere pressure.

THE INTERNAL ENERGY

We now ask ourselves this important question: If a change occurs in which heat energy is liberated, such as the oxidation of magnesium just discussed, where does this energy come from? Alternatively, if a change occurs in which heat energy is absorbed by a system, such as the vaporization of water, where does this energy go? Such questions are not trivial. The Law of Conservation of Energy tells us that energy cannot be created or destroyed. Any appearance of energy in one place can only be balanced by its disappearance elsewhere.

The answers to our questions are easily given in terms of the energy levels discussed in Chapters 2 and 3. When a system loses heat energy, particles move from levels of higher energy to levels of lower energy. Conversely, if a system gains heat energy, particles move from lower to higher levels. Two idealized examples may help to make this point clearer.

In Figure 5-1 we illustrate what happens when a substance is heated without any chemical change occurring. We take our substance to be a crystalline element in which only vibrational motion is possible. The levels shown, which are 500 cal mole^{-1} apart, correspond to vibrational motion in the x-direction. The figure refers to a mole of atoms. The population of each level is indicated, as before, by the thickness of the line and also as a fraction of a mole.

We can easily see in this figure that as the temperature increases, levels of higher energy become populated at the expense of lower lying levels. This means that the total energy content of the system increases with temperature. The total energy of the system, i.e., the sum of the individual energies of all the particles in the system, is given a special symbol and a special name. It is called the *internal energy* and is specified by the symbol E. In quantitative terms E is given

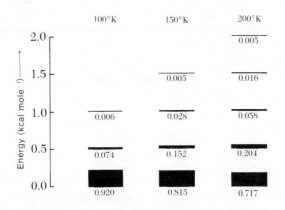

Figure 5-1. As the temperature rises, higher levels become more populated and the internal energy of the crystal increases.

by the formula:

$$E = \sum_{i=0}^{i=\infty} n_i\epsilon_i = n_0\epsilon_0 + n_1\epsilon_1 + n_2\epsilon_2 + n_3\epsilon_3 + \ldots \tag{5.3}$$

The products of the population $(n_0, n_1, n_2 \ldots)$ and the energy $(\epsilon_0, \epsilon_1, \epsilon_2 \ldots)$ for each level are summed for all levels beginning at level 0 and continuing through the highest occupied level.

As an example, we now calculate the internal energy of the mole of atoms in Figure 5-1 at $100°K$. We take the ground state as having zero energy. This means that the total energy of the 0.920 mole of atoms in this level is by definition zero. The first excited state is 500 cal mole^{-1} in energy above the ground state and has a population of 0.074 mole of atoms. The total energy of these atoms is thus 0.074 mole $\times$ 500 cal mole^{-1} = 37.0 cal. A similar calculation shows that the total energy of the atoms in the second excited state is 0.006 $\times$ 1000 = 6 cal. The internal energy, E, of the whole system is thus 0 + 37.0 + 6.0 = 43 cal at $100°K$.

Using the same method, we find that at $150°K$

$$E_{150} = (0.815 \times 0) + (0.152 \times 500) + (0.028 \times 1000) + (0.005 \times 1500)$$

$$= 112 \text{ cal}$$

Again, at $200°K$, E_{200} = 194 cal. From these calculations, we see that the internal energy rises with temperature as we would expect.

A second example of an energy change in a system is shown in Figure 5-2. This figure indicates what happens, in terms of energy levels and populations, when a chemical reaction of the type

$$A \rightarrow B \quad (200°K) \tag{5.4}$$

occurs. The reaction can be thought of as the conversion of isomer A into isomer B under conditions of constant temperature.

In this figure, the left-hand set of levels corresponds to compound A and the right-hand set to compound B. The energy levels of A are evenly spaced 500 cal mole^{-1} apart, while those of B are spaced 200 cal mole^{-1} apart. The gound state of A is exactly 1500 cal mole^{-1} *higher* in energy than the ground state of B.

The two Boltzmann distributions shown in the figure are computed for one mole of molecules at $200°K$. That given for A is for a mole of pure A and represents the Boltzmann distribution *before* reaction (5.4) begins. Similarly, that given for B represents the Boltzmann distribution *after* the reaction has gone to *completion*. It is clear that the reaction results in a *decrease* in the internal energy E. The molecules of B end up occupying levels almost all of which are

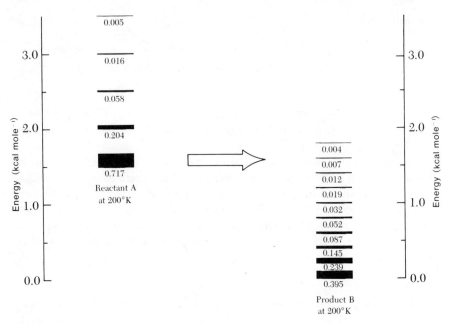

Figure 5-2. Diagrammatic representation of an exothermic chemical reaction in terms of energy levels.

lower in energy than the ground state of compound A.

We can use the populations of the levels given in Figure 5-2 to calculate the internal energy of one mole of A and the internal energy of one mole of B. The difference corresponds to the loss in internal energy. We will take the lowest energy level in the figure, the ground state of B, as our zero of energy. Using the method employed in the previous example, we find that the internal energy, E_A, of one mole of A at 200°K is (1500 + 194) cal and that of one mole of B, E_B, is 290 cal. The loss in internal energy accompanying the reaction is thus (1500 + 194) – 290 = 1404 cal.

What happens to this 1404 cal of energy when the reaction A →B occurs? It cannot simply disappear; the law of conservation of energy forbids such behavior! The loss in internal energy of the reaction system appears as heat energy. The reaction A →B is an *exothermic reaction*. As A changes into B, heat is liberated and the temperature of the surroundings rises.

We could have deduced that this reaction is exothermic by simply noting that *the ground state of B is lower in energy than the ground state of A*. The loss in internal energy is 1404 cal. This figure is not very different from the 1500 cal mole⁻¹ energy difference between the ground states of A and B. In other words, we could have obtained a good approximation for the loss in internal energy by assuming all of the molecules to be in the ground state both before and after the reaction. Although some 30 per cent of the molecules of A are excited before the reaction begins and some 60 per cent of the molecules of B are excited after the reaction is complete, this makes very little difference to the total energy change. To a large extent the energy involved in exciting the molecules above the ground state in the two cases (194 cal and 290 cal) cancel each other out.

The behavior just described is not confined to the example of Figure 5-2, but is typical of almost all chemical reactions. When a chemical reaction occurs there is always a change in bonding, i.e., a change in *electronic* energy. Changes in electronic energy are usually much larger in magnitude than changes in other forms of molecular energy. Because of this, we can almost always approximate the change in internal energy that accompanies a chemical reaction by calculating the change in *electronic energy*, neglecting changes in vibrational, rotational, and translational energy.

Such an approximation gives a result which is seldom in error by more than a few percent.

THE ZERO OF ENERGY

In both examples considered above, we took the lowest energy level as our zero of energy. This was purely a matter of convenience. The lowest level does not actually correspond to a situation of zero energy (i.e., to the atoms having no potential or kinetic energy at all). In fact, it is doubtful whether the concept of a true zero of energy has any useful meaning. A body with zero kinetic energy for instance, would have to be stationary, but it is difficult to define exactly what we mean by a stationary body. The objects we usually regard as stationary are in reality revolving around the earth's axis, which in turn is moving around the sun. The sun in turn moves round the center of the galaxy — and so on. There is no obvious reference point which we can say is not moving. A similar difficulty intrudes when we try to establish a zero for potential energy. Do we take the desk, the floor, sea level, the center of the earth, or some other point in the universe as our zero for potential energy?

The best we can do in the face of such difficulties is to select an arbitrary but convenient reference state for the internal energy, and ascribe zero energy to it. We do essentially the same thing when we talk about altitude in terms of height above sea level, which is taken as the arbitrary zero for altitude. Furthermore, we lose nothing by setting an arbitrary zero for energy. Invariably in chemistry, we are only interested in *differences* in energy, never in their absolute values.

Students often suffer from a misconception about zero energy. They think that all systems automatically have zero internal energy at the absolute zero of temperature because "everything stops moving" at this temperature. A moment's reflection will reveal that this cannot be the case. If electrons, for instance, stopped moving about their nuclei, there would be an enormous reduction in the volume of all substances at absolute zero! Such a phenomenon has not yet been observed.*

It is important to realize that all compounds do *not* have the same energy at the absolute zero of temperature. As an example of this, consider the two substances A and B shown in Figure 5-2. If we lower the temperature of a mole of A to $0°K$, only the ground level of A will be occupied. Similarly, if we lower the temperature of a mole of B to $0°K$, this will result in only the ground level of B being occupied. Since the difference in ground levels is 1500 cal mole^{-1}, this means that even at absolute zero the internal energy of a mole of A is 1500 cal higher than that of a mole of B.

MICROSCOPIC AND MACROSCOPIC VIEWS OF MATTER

Up to this point in the book, we have for the most part, taken what is called the *microscopic* view of matter. We have regarded matter as being made up of atoms and molecules occupying translational, rotational, and vibrational levels. From now on we will begin to take another, more familiar view of matter, the *macroscopic* view. In this second view of matter we treat it as we normally experience it, in bulk.

We have already encountered the difficulty which forces us to change from the microscopic to the macroscopic view of matter. It is the sheer complexity of regarding matter as a myriad of competing energy levels. Although the microscopic view has enabled us to see what is really at stake in chemical equilibrium, it can be applied quantitatively only to the very simplest cases of chemical equilibrium. It is possible, though, to correlate the microscopic view of matter with *macroscopic* properties. For instance, the difference in internal energy between reactants and products which we interpreted in terms of energy levels in Figure 5-2 can be linked to the heat changes that accompany the reaction. Once we have made such a connection, even complex cases of chemical equilibrium can be handled.

THE STATE OF A SYSTEM

What we normally call the macroscopic properties of a system can be conveniently divided into two kinds: intensive and extensive properties. The *intensive* properties of a system are those properties such as the pressure, the temperature, and the composition of the system, which are independent of the quantity of matter in the system. The *extensive* properties of a system, by contrast, are those which are directly proportional to the quantity of matter in the system. Examples of this kind of property are the mass and the volume. If

---

*Even more fundamentally, the "everything stops" fallacy contradicts the uncertainty principle.

we double the number of particles of each kind without altering its intensive properties, we also double the mass and the volume of the system.

As an example of a system whose properties we can discuss, let us take 100 ml of 1M NaCl solution at 1 atmosphere pressure and at 25°C. Experience tells us that this system will have properties that are *identical* to those of a second 100 ml of 1M NaCl solution at the same temperature and pressure. Not only will the volume, the composition, the temperature and pressure be the same (we have just said that they are); the mass, the density, and the refractive index of both systems will also be identical. In short, *all* of the properties of the two systems, both intensive and extensive, will be identical. Under such circumstances we say that the two systems are in the *same state*. If some or all of the properties of one system change, then we say that the *state of that system* has changed. The two systems will then no longer have identical properties, and hence will no longer be in the same state.

In order to characterize or *define* the state of a system we do not have to specify *all* the properties of the system. A limited number will do. For the sodium chloride solution, for instance, it is only necessary to specify:

1. Pressure (1 atmosphere)
2. Temperature (25°C)
3. Volume (100 ml)
4. Composition (1 mole liter^{-1})

Any two samples of NaCl solution for which all of these four properties are identical *must necessarily be in the same state*.

The choice of properties to be specified in defining the state of a system is somewhat arbitrary. In the above case, for instance, the mass of the NaCl solution would have done just as well as the volume. The *number* of properties to be specified, however, is *not* arbitrary. To define the state of a solution of NaCl, the magnitudes of *four* properties, one of them extensive, must be given.*

What we have so far called the properties of a system are often called "functions of the state of the system" or, more simply, *state functions*. A state function is one that depends only on the state of the system, and not on anything else. When the state of a system changes, some or all of the state functions will change in value. If the state of a system does *not* change, *none* of the state functions will change in value.

It is fairly obvious that the internal energy E defined by equation (5.3) must be a state function. Consider any two systems that are in the same state, such as the two 100 ml samples of NaCl solution just considered. Both systems will have identical sets of energy levels with identical populations; i.e., both will have identical Boltzmann distributions. The sum $\Sigma n_i \epsilon_i$ of equation (5.3) will then be the same for both systems. Thus, if two systems are in the same state they have the same internal energy.

---

*Some authors prefer to regard two systems as being in the same state if only the *intensive properties* are identical. In such a convention, a 50 ml sample of 1M NaCl at 25°C and one atmosphere is in the same state as the 100 ml samples considered above.

E is also an *extensive property* of a system. Again this follows directly from the definition, equation (5.3):

$$E = \Sigma n_i \epsilon_i$$

If we double the total number of particles of each kind in a system without altering the distribution among energy levels, we will simply double the population of every level and hence double the energy of the system.

CHANGES IN THE STATE OF A SYSTEM

When a system changes from one state to another, some or all of its properties will change. To describe such changes, the initial state of the system is indicated by the symbol 1 and the final state by the symbol 2. Furthermore, if a property of the system, which we will call X, changes from an initial value of X_1 to a final value of X_2, the symbol ΔX is used to indicate the change in the value of the property. Thus, ΔX is defined by the equation

$$\Delta X = X_2 - X_1$$

Let us consider the changes in macroscopic properties that occur when one mole of liquid water is converted into one mole of gaseous water at 100°C and 1 atmosphere pressure:*

$$H_2O(l) \rightarrow H_2O(g) \quad (100°C, 1 \text{ atm}) \tag{5.5}$$

For the purpose of the argument, we imagine this change to be carried out in the apparatus shown in Figure 5-3, where the water is enclosed in a container fitted with a tight, but frictionless, piston.

When reaction (5.5) occurs, the volume expands from 18 ml (the volume of 18 g of liquid water) to 30.6 liters (the molar volume of a gas at 100°C). At the same time, 9.72 kcal of heat are absorbed from the surroundings. We can thus write

$$V_1 = 0.018 \text{ liter}$$

and

$$V_2 = 30.6 \text{ liter}$$

---

*Note that such an equation refers not only to the substances involved in the change but also to the *amount*. Specifically, equation (5.5) refers to *one mole* of H_2O. If two moles were involved, we should have written

$$2 H_2O(l) \rightarrow 2 H_2O(g) \quad (100°C, 1 \text{ atm})$$

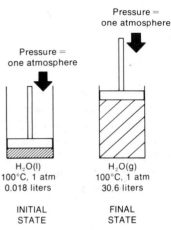

Figure 5-3. Some of the energy absorbed in vaporization of water is used to do expansion work against the pressure of the atmosphere.

so that

$$\Delta V = V_2 - V_1$$

$$= 30.6 \text{ liter} - 0.018 \text{ liter}$$

$$= 30.6 \text{ liter}$$

What is ΔP for this change? Since $P_1 = P_2 = 1$ atm, $\Delta P = P_2 - P_1 = 0$. Similarly, ΔT is also zero. We come now to the $64,000 question. What is the change in the internal energy of the mole of H_2O? What is ΔE for reaction (5.5)?

At first sight it might appear that ΔE was given by the 9.72 kcal of heat energy absorbed in the reaction. Although this is approximately true, it is not exactly so. There is a further quantity of energy we must still account for, the so-called "expansion work." The heat energy absorbed by the H_2O does not all go to raise H_2O molecules to higher energy levels. Some of it is used to drive the piston shown in Figure 5-3 upwards against the force exerted by the atmosphere. Before we can calculate ΔE, we must know the magnitude of this expansion work.

We shall deal more fully with the subject of expansion work in Chapter 8, where we shall prove that for an expansion occurring at constant pressure, the work done by the system is $P\Delta V$. For the moment, let us merely note that this quantity has the dimensions of energy. P is a measure of force per unit area, so that it has the dimensions force $\times$ length^{-2}. The product $P\Delta V$ thus has the dimensions (force $\times$ length^{-2}) (length^3), or force $\times$ length. These are the same dimensions as those of energy.

Since $P = 1$ atm and $V = 30.6$ liters, the product $P\Delta V$ has the value 30.6 liter-atmospheres. Now,

1 liter-atmosphere = 24.2 cal

so that

$$P\Delta V = 30.6 \text{ lit-atm} \times \frac{24.2 \text{ cal}}{1 \text{ lit-atm}} = 741 \text{ cal}$$

The work done when one mole of H_2O expands in volume from 18 ml to 30.6 liters at a pressure of one atmosphere is thus 741 cal.

We can now calculate ΔE. The 9.72 kcal of heat energy absorbed by a mole of water in the process of vaporization is used up doing two things: (i) raising H_2O molecules from lower to higher energy levels, involving a total amount of energy ΔE, and (ii) doing expansion work (of magnitude 741 cal) against the atmosphere. We can thus write:

$$9.72 \text{ kcal} = \Delta E + 0.74 \text{ kcal}$$

or

$$\Delta E = 8.98 \text{ kcal}$$

A change in internal energy ΔE is thus not only something that we can interpret on the microscopic level, but also something that we can calculate from experimentally observed changes in the state of a system. In fact we can find ΔE experimentally, without knowing anything at all about the details of the energy levels of the system we are studying.

THE FIRST LAW OF THERMODYNAMICS

The method we have used to calculate ΔE in the case just considered can in general be applied to *any* change which *any* system undergoes. Let q be the amount of heat energy *absorbed* by a system and let w be the amount of work done *by* the system during such a change. We can then, as we did previously, regard the heat energy absorbed as being used up in (i) raising the molecules from levels of lower energy to levels of higher energy and (ii) doing external work. Accordingly, we can write.

$$q = \Delta E + w$$

or

$$\Delta E = q - w \qquad\qquad \textbf{(5.6)}$$

Equation (5.6) is an important relationship, important enough to rate a special name of its own. It is called the *First Law of Thermodynamics*. The First Law is really a restatement of the Law of Conservation of Energy in terms of the internal energy of the system.

In order to familiarize ourselves with the First Law and the sign conventions as they pertain to heat absorbed and work done, let us consider an example of its operation. The example we shall take is that of the lead-copper voltaic cell shown in Figure 5-4. In the left-hand compartment is a 1M solution of $Pb(NO_3)_2$ surrounding an electrode of pure lead. In the right-hand compartment, similarly, is a 1M solution of $Cu(NO_3)_2$ surrounding a copper electrode. The two compartments are separated by a porous barrier which prevents the two solutions from mixing with each other, but allows ions to conduct charge across the barrier when the cell is in operation.

When the lead and copper electrodes are connected electrically, the following processes occur at the electrodes:

At the left-hand electrode:

$$Pb \rightarrow Pb^{2+}(1M) + 2e \qquad (5.7a)$$

At the right-hand electrode:

$$Cu^{2+}(1M) + 2e \rightarrow Cu \qquad (5.7b)$$

The net chemical change which occurs is the reaction:

$$Pb + Cu^{2+}(1M) \rightarrow Cu + Pb^{2+}(1M) \qquad (25°C, 1 \text{ atm}) \qquad (5.7)$$

In other words, the lead metal reduces the copper(II) ion to metallic copper.

We consider the initial state of our system to be that of Figure 5-4 at 25°C and 1 atmosphere, and the final state to be the same cell at the same temperature and pressure when 1 mole of Pb has been oxidized and 1 mole of Cu^{2+} has been reduced without any significant changes in the concentrations of the solutions involved having occurred. It is possible to allow the cell to proceed from this initial state to this final state in a variety of ways. In general, we find that the values of both q and w vary depending on how the process is carried out, but that the difference q − w is independent of path.

Let us now examine two different ways of effecting the change from the

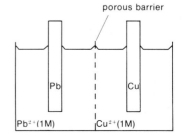

Figure 5-4. A lead-copper voltaic cell.

initial to the final state of the cell, paying particular attention to the values of q and w in each case. In so doing, we will ignore the expansion work $P\Delta V$, since the volume change for reaction (5.7) is extremely small. We will also ignore the energy changes which occur due to the ions migrating through the porous barrier.

(1) In our first example we allow the cell to drive an electric motor, which operates a pulley and lifts a weight, thus performing mechanical work. If the motor were 100 per cent efficient, we should find that the mechanical work done would be 17.8 kilocalories.

Accordingly we can write:

$$w = 17.8 \text{ kcal}$$

Notice that we observe the convention that w is *positive* if the system is *doing* work.

At the same time as the change detailed above is occurring, the voltaic cell will be found to *absorb* 5.8 kcal of heat from its surroundings. We can accordingly write

$$q = 5.8 \text{ kcal}$$

observing the convention that heat *absorbed* by the system is regarded as *positive*.

Applying the First Law, we find $\Delta E = q - w = 5.8 - 17.8 = -12.0$ kcal. The voltaic cell has lost internal energy, since it has performed more work than can be balanced by the absorption of heat energy.

(2) We could also allow our lead-copper cell to proceed from the same initial to the same final state in a somewhat different manner by short-circuiting the cell with a thick copper wire of very low resistance. Lacking an electrical motor to operate, the electrons flowing from one electrode to the other through the thick copper wire would perform no work at all. Because no work would be done, we can write

$$w = 0.0 \text{ kcal}$$

If the short-circuiting wire had zero resistance, no heating would occur in it. The cell itself, however, would *release* 12.0 kcal of heat energy to its surroundings. Thus:

$$q = -12.0 \text{ kcal}$$

The *negative* sign, by convention, indicates that the system has *released* rather than absorbed heat.

Again we can apply the First Law, and find that

$$\Delta E = q - w = -12.00 - 0.0$$

$$= -12.0 \text{ kcal}$$

which is the same result as previously.

That we obtain the same result for ΔE in both cases is exactly what we would expect. The internal energy is a *state function*, its value depending only on the state of the system. A change in the internal energy, ΔE, will depend only on the nature of the initial and the final states (i.e., on their energy levels) and will not depend in the least on *how* we effect the change from one state to another. That ΔE is independent of path in this way is a very important aspect of its behavior which we shall soon use to a very good purpose.

The First Law suggests how we might measure ΔE. Since

$$\Delta E = q - w$$

it follows that if we measure q, the heat absorbed during a change in state, and w, the work done by the system during the same change, we can calculate ΔE by taking the difference between these two quantities. We will discuss in Chapter 8 how such heat and work quantities may be measured. For present purposes we will assume that such measurements can be made. We should, however, observe two points in this connection.

In the first place, note that such measurements pertain to the bulk behavior of the system and do not require any information concerning the population of energy levels. In other words, we are able to determine values for ΔE *without having to know anything about the structure of our system on the microscopic level.*

The second point to keep in mind is that w, the work produced, is often a small quantity. In most laboratory situations, the system in which we are interested is not harnessed to do electrical work. In most reactions that we carry out in the laboratory, the only work done is expansion work. Furthermore, this expansion work is often very small. In a typical case, we might mix a solution of an acid with a solution of a base. In such a case the change in volume is minute and the expansion work quite negligible. In Table 5-1 we give a list of ΔE values and expansion work values for several reactions. We can see that the expansion work is an appreciable fraction of ΔE only when ΔE is small (reaction 2) or when a large change in volume occurs due to the production or the disappearance of several moles of gas (reaction 4). Even in these two cases, w is less than 10 per cent of ΔE.

For many purposes when high accuracy is not required, we can take ΔE as equal to q. The error involved is seldom more than a few per cent.

Table 5-1

Values for ΔE and expansion work for several chemical reactions
at 25°C and a constant pressure of one atmosphere.

Reaction	ΔE kcal	$w = P\Delta V$ kcal	$w/\Delta E$
(1) $Mg(s) + \frac{1}{2} O_2 \rightarrow MgO$	-143.5	-0.30	0.002
(2) $H_2O(l) \rightarrow H_2O(g)$	$+9.9$	$+0.59$	0.06
(3) $H_2(g) + Cl_2(g) \rightarrow 2\ HCl(g)$	-44.2	very small	$0.0000\ldots$
(4) $2H_2(g) + O_2(g) \rightarrow 2\ H_2O(l)$	-134.8	-1.78	0.013
(5) $CuO(s) + H_2O(l) \rightarrow Cu(OH)_2(s)$	-15.6	-4.11×10^{-5}	0.000003

HESS'S LAW

We saw above that if a system undergoes a change from state 1 to state 2, the change in internal energy, ΔE, will depend only on the nature of states 1 and 2, and not on any details of the path by which the change was effected. As an illustration of what this statement means, let us consider two different ways of oxidizing graphite to carbon dioxide. We could oxidize two moles of graphite to carbon dioxide directly in one reaction:

$$2\ C(graphite) + 2\ O_2(g) \rightarrow 2\ CO_2(g) \quad (25°C, 1\ atm) \qquad \textbf{(5.8)}$$

or we could first oxidize them to carbon monoxide:

$$2\ C(graphite) + O_2(g) \rightarrow 2\ CO(g) \quad (25°C, 1\ atm) \qquad \textbf{(5.9)}$$

and then oxidize the monoxide to the dioxide:

$$2\ CO(g) + O_2(g) \rightarrow 2\ CO_2(g) \quad (25°C, 1\ atm) \qquad \textbf{(5.10)}$$

Whether we choose the one-stage path of reaction (5.8) or the two-stage path of reactions (5.9) and (5.10), the initial and final states are the same: 2 moles of graphite and two moles of O_2, both at 25°C and 1 atmosphere, on the one hand, and two moles of CO_2 at the same temperature and pressure on the other. The internal energy change for both paths must thus be the same; i.e., ΔE for reaction (5.8) must be equal to the sum of ΔE for reaction (5.9) and ΔE for reaction (5.10). The values of ΔE for these three reactions are shown below.

$2C(graphite) + O_2(g) \rightarrow 2CO(g)$	-53.37 kcal	(5.9)
$2CO(g) \quad + \quad O_2(g) \rightarrow 2CO_2(g)$	-134.69 kcal	(5.10)
$2C(graphite) + 2O_2(g) \rightarrow 2CO_2(g)$	-188.06 kcal	(5.8)

Notice that we can cancel the two moles of CO on the right-hand side of equation (5.9) with two moles of CO on the left-hand side of equation (5.10), since all the CO produced at the half-way stage of the two-stage path is subsequently oxidized to CO_2. Equations (5.9) and (5.10) can thus be added as though they were algebraic equations to give the net process of equation (5.8).

The general principle that ΔE values for several reactions can be added together in this way to yield the ΔE value for the net reaction is sometimes referred to as *Hess's Law*. Using this law, it is possible to add and subtract values of ΔE for reactions that have already been studied in order to obtain the value of ΔE for a reaction that has not been studied or that cannot be studied directly.

An example of the use of Hess's Law to find the value of ΔE for a reaction that is difficult to study experimentally is the calculation of ΔE for the reaction

$$2\,C(\text{graphite}) + H_2(g) \rightarrow C_2H_2(g) \quad (25^\circ C, 1\,\text{atm}) \qquad \textbf{(5.11)}$$

The gas acetylene, C_2H_2, cannot be prepared directly by this reaction. Nevertheless, it is possible to calculate ΔE for reaction (5.11) from the following values of ΔE, all quoted for $25^\circ C$ and 1 atmosphere.

Reaction	$\Delta E(kcal)$	
$C(\text{graphite}) + O_2(g) \rightarrow CO_2(g)$	-94.1	**(5.12)**
$2\,H_2(g) + O_2(g) \rightarrow 2\,H_2O(l)$	-134.8	**(5.13)**
$C_2H_2(g) + \frac{5}{2}O_2(g) \rightarrow 2\,CO_2(g) + H_2O(l)$	-311.1	**(5.14)**

To obtain equation (5.11), we first multiply equation (5.12) by two and equation (5.13) by one half, and add the two results. We next subtract equation (5.14) by writing it down in reverse and adding. The ΔE values are manipulated in exactly the same way as the equations. The results of these manipulations are indicated below:

Reaction	$\Delta E(kcal)$	
$2\,C(\text{graphite}) + 2\,O_2(g) \rightarrow 2\,CO_2(g)$	$2\,(-94.1)$	$= -188.2$
$H_2(g) + \frac{1}{2}O_2(g) \rightarrow H_2O(l)$	$\frac{1}{2}\,(-134.8)$	$= -67.4$
$2\,CO_2(g) + H_2O(l) \rightarrow C_2H_2(g) + \frac{5}{2}O_2(g)$	$-(-311.1)$	$= +311.1$
$2\,C(\text{graphite}) + H_2(g) \rightarrow C_2H_2(g)$		$+55.5$

We thus obtain the final result that ΔE for reaction (5.11) is $+55.5$ kcal.

THE ENERGIES OF COVALENT BONDS

The value of ΔE for most chemical reactions can be found either by direct measurement of the heat changes and volume changes that accompany the reaction, or indirectly by the use of Hess's Law. Although such information is useful, it gives little or no insight into what is happening on the molecular level. Careful measurements, for instance, on the reaction

$$H_2(g) + Cl_2(g) \rightarrow 2HCl(g) \quad (25°C, 1 \text{ atm}) \tag{5.15}$$

tell us that ΔE is -44.214 kcal. Although such a value is very exact, it gives us no explanation of *why* this reaction is exothermic rather than endothermic.

There is no simple general answer to the question of why a particular reaction is exothermic or endothermic. In general, the sign of ΔE will depend on whether the reactants and products are solids, liquids, or gases, and whether the bonding involved is ionic, covalent, or metallic. A complete discussion of the subject is outside the scope of an introductory text. The best we can do here is to deal with the simplest type of reaction, one in which both reactants and products are gases. It is quite easy to explain what determines the magnitude of ΔE for a gaseous reaction because only one kind of bonding, the covalent type, is involved. In addition, the effect of intermolecular attraction can be neglected.

The best way of looking at the energy changes that accompany a gaseous reaction is in terms of a hypothetical two-stage process. In the first stage the reactant molecules are regarded as being broken up into their constituent atoms. In the second stage these atoms are reconstituted into molecules of the products. From this point of view the reaction becomes a succession of bond-breaking and bond-making processes. The change in internal energy is then seen to be governed by the energies involved in making and breaking covalent bonds.

We have already dealt with the forces and energies involved in the covalent bond in Chapter 3. We saw there how energy must be expended in order to pull the two nuclei involved in a covalent bond away from each other. As shown in Figure 3-7, if enough energy is used it is possible to separate the two nuclei entirely, resulting in the complete dissociation of the molecule.

The energy required to dissociate a diatomic molecule is usually called its *bond dissociation energy*. We shall define the bond dissociation energy of the molecule XY or the bond X—Y as the energy necessary, per mole of XY molecules, for the process

$$XY(g) \rightarrow X(g) + Y(g)$$

and give it the symbol D_{XY}. Thus, for example, the bond dissociation energy of the H—H bond, D_{HH}, is the energy required to dissociate a mole of hydrogen molecules according to the reaction

$$H_2(g) \rightarrow 2 H(g)$$

The value of this bond dissociation energy is found from studies of the spectrum of H_2 to be 104 kcal mole^{-1}. Another example is the bond dissociation energy of the HI molecule, D_{HI}. This has a value of 71 kcal mole^{-1}, which tells us that it requires 71 kcal of energy to dissociate a mole of HI molecules into atoms.

We have purposely left the definition of a bond dissociation energy a little vague. Should we take D_{XY} as being the energy change in the following process:

$$\left. \begin{array}{l} \text{One mole of XY molecules} \\ \text{all in the ground state} \end{array} \right\} \rightarrow \left\{ \begin{array}{l} \text{One mole stationary X atoms} \\ + \text{ One mole stationary Y atoms} \end{array} \right.$$

or should we take account of the fact that atoms can have translational energy and molecules can have translational, rotational, and vibrational energies? Should we not perhaps take D_{XY} as being equal to ΔE for the reaction

$$XY(g) \rightarrow X(g) + Y(g) \qquad (1 \text{ atm})$$

at 25°C or some other temperature?

In fact, such nice distinctions mean very little in terms of energy. For H_2 the ground level-to-ground level dissociation energy is 103.257 kcal mole^{-1}, while the ΔE value for the dissociation at 25°C and one atmosphere is 103.59 kcal mole^{-1}. This is a good example of the rule stated previously, that the magnitude of ΔE is determined mainly by the change in electronic energy, and that changes in vibrational, rotational, and translational energy make only a small contribution.

In what follows, we shall use the term bond energy interchangeably to mean either the ground state difference in energies or the value of ΔE at 25°C or any other temperature. We shall ignore any small differences in energy that there may be between these various alternatives.

We should notice that when we talk about the bond dissociation energy the "bond" to which we refer is *not* always a single bond. For instance, when we talk about breaking the nitrogen-nitrogen bond in the molecule N_2, we are referring to the process:

$$N\equiv N(g) \rightarrow 2N(g)$$

and thus to the breaking of a *triple bond*. In this book, if we wish to draw particular attention to the fact that a bond is a multiple bond, we shall indicate it in the conventional way; e.g., $O=O$ or $N\equiv N$. On the other hand, we will often want to refer to the breaking of a bond, say the X–Y bond, irrespective of whether it is multiple or not. To indicate such a sense of the word "bond" we shall use the symbol X–Y rather than distinguish among X–Y, X=Y, and X≡Y.

Table 5-2 gives the values of bond dissociation energies for some diatomic molecules, deduced mainly from their spectra. We can use these values to predict ΔE for any gaseous reaction in which only the molecules listed in the table are involved. To do this we must look at the reaction as a succession of bond-breaking and bond-making processes.

Table 5-2

Bond dissociation energies for some diatomic molecules (kcal mole^{-1}).

N_2	226	NO	151
O_2	119	CO	257
H_2	104	HF	136
F_2	38	HCl	103
Cl_2	58	HBr	88
Br_2	46	HI	71
I_2	36		

As an example, consider the reaction

$$H_2(g) + F_2(g) \rightarrow 2HF(g) \quad (25°C, 1 \text{ atm}) \quad \textbf{(5.16)}$$

We can regard this reaction as the sum of the following reactions, each of which corresponds either to the making or the breaking of a single type of bond.

Reaction	*ΔE (kcal)*
$H_2(g) \rightarrow 2 H(g)$	104
$F_2(g) \rightarrow 2 F(g)$	38
$2 H(g) + 2 F(g) \rightarrow 2 HF(g)$	$-2(136)$

The value of ΔE for each of these processes is easily deduced from Table 5-2 if we remember that a bond-breaking process results in an *increase* in the internal energy, while a bond-making process results in a *decrease* in the internal energy. Adding the three values together in accordance with Hess's Law gives the value of ΔE for the net process, reaction (5.16). The final result is thus

$$\Delta E = D_{HH} + D_{FF} - 2D_{HF}$$

$$= 104 + 38 - 2 \times 136$$

$$= -130 \text{ kcal}$$

We are now in a position to explain *why* it is that the reaction between hydrogen and fluorine is an exothermic one. It is because the bonds that have been *broken* are *weaker,* on average, than the bonds that have been *made.* This is best seen in a diagram such as Figure 5-5. The two bonds that are broken are the rather weak F—F bond (38 kcal mole^{-1}) and the fairly strong H—H bond (104

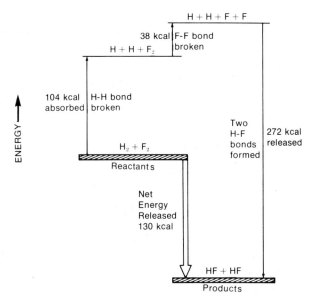

Figure 5-5. Energy diagram for formation of hydrogen fluoride from hydrogen and fluorine molecules.

kcal mole^{-1}), while the two bonds that are made are both strong H—F bonds (136 kcal mole^{-1}). Since one weak and one fairly strong bond are broken while two very strong bonds are formed, the net result is a reduction in the internal energy and the release of heat.

We can see from this particular example the general principle that determines whether a gaseous reaction will be exothermic or endothermic. If it requires less energy to break the reactant molecules into their constituent atoms than is released when these atoms are reconstituted into product molecules, then the reaction will be exothermic. If the opposite is true, then the reaction will be endothermic.

MEAN BOND ENERGIES

While the bond dissociation energy in a *diatomic* molecule is an unambiguous quantity, in a *polyatomic* molecule the meaning is less clear. For example, it is found that ΔE is +113.4 kcal for the reaction

$$H—O—H(g) \rightarrow H(g) + O—H(g) \quad (25°C, 1 \text{ atm})$$

in which one of the O—H bonds of water is broken. On the other hand, ΔE = +100.6 kcal for the reaction

$$O—H(g) \rightarrow O(g) + H(g) \quad (25°C, 1 \text{ atm})$$

in which the second O–H bond is broken. Which of these two values are we to take as a measure of the strength of the O–H bond, or should we perhaps take the average?

The position is further complicated when we look at other compounds containing O–H bonds. It appears that a slightly different amount of energy is required to break an O–H bond in each case. For instance, $\Delta E = +102$ kcal for the reaction

$$CH_3O\text{–}H(g) \rightarrow CH_3O(g) + H(g) \quad (25°C, 1 \text{ atm})$$

while $\Delta E = +90$ kcal for the reaction

$$H\text{–}O\text{–}O\text{–}H(g) \rightarrow H\text{–}O\text{–}O(g) + H(g) \quad (25°C, 1 \text{ atm})$$

Because of this difficulty, an average value for the O–H bond energy, taken over a fair number of compounds, is usually quoted. Such a value of the bond energy is called a *mean bond energy*. The mean bond energy for the O–H bond is most often quoted as 111 kcal mole^{-1}, although the value varies with the source. Such a mean bond energy is usually reliable to within 10 kcal mole^{-1}, although it can sometimes be as much as 20 kcal mole^{-1} in error in a particular case.

USE OF BOND ENERGY TABLES

A set of mean bond energies is given in Table 5-3 and also inside the back cover. We can use such a table of bond energies to calculate ΔE for a large number of gaseous reactions. The method used is the same as that employed previously for diatomic molecules using Table 5-2. The reaction is broken up into two hypothetical stages: a bond-breaking stage and a bond-making stage.

Example: Use the bond energies given in Table 5-3 to calculate ΔE for the following reaction:

$$CH_4(g) + 4 F_2(g) \rightarrow CF_4(g) + 4 HF(g) \tag{5.17}$$

We divide the reaction up into the usual two stages. In the first stage all the bonds of the reactant molecules are broken to give only atoms:

$$CH_4(g) + 4 F_2(g) \rightarrow C(g) + 4 H(g) + 8 F(g) \tag{I}$$

Since four C–H bonds and four F–F bonds must be broken, the internal energy is increased by an amount equal to four times the C–H bond energy plus four times the F–F bond energy. In other words:

$$\Delta E_I = 4D_{C\text{–}H} + 4D_{F\text{–}F} = 4 \times 99 \text{ kcal} + 4 \times 37 \text{ kcal} = 544 \text{ kcal}$$
(See Figure 5-6 for an energy diagram.)

Table 5-3

Mean Bond Energies (expressed in kcal per mole of bonds)

				Single Bonds					
	H	**C**	**N**	**O**	**F**	**S**	**Cl**	**Br**	**I**
I	71	57	–	48	–	–	50	43	36
Br	88	66	58	–	57	51	52	46	
Cl	103	79	48	49	61	66	58		
S	81	62	–	–	68	63			
F	135	116	65	44	37				
O	111	82	48	33					
N	93	70	38						
C	99	83							
H	104								

Multiple Bonds

$C=C$	147	$C=N$	147	$C=O$*	192	$N=N$	100	
$C\equiv C$	200	$C\equiv N$	210	$C\equiv O$	256	$N\equiv N$	226	
$O=O$	119	$S=S$	100	$S=O$	125			

*This is the value for CO_2.

In the second stage the atoms produced in Stage I are recombined into molecules of product:

$$C(g) + 4 H(g) + 8 F(g) \rightarrow CF_4(g) + 4 HF(g) \qquad \textbf{(II)}$$

This corresponds to the *formation* of four C–F bonds and four H–F bonds (see Figure 5-6), with a consequent *lowering* of the internal energy. Thus

$$\Delta E_{II} = -(4D_{C-F} + 4D_{H-F}) = -(4 \times 116 \text{ kcal} + 4 \times 135 \text{ kcal}) = -1004 \text{ kcal}$$

Adding the two internal energy changes, we find that

$$\Delta E = \Delta E_I + \Delta E_{II} = 544 - 1004 = -460 \text{ kcal}$$

The exact result derived from experiment is −462 kcal. We do not expect an

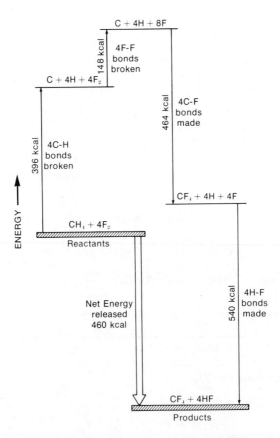

Figure 5-6. Energy diagram for formation of carbon tetrafluoride and hydrogen fluoride from methane and fluorine.

exact answer because of the use of *mean* bond energies for the C–H and C–F bonds.

It is possible to derive a general formula for calculating ΔE for any gaseous reaction from a table of bond energies. Again we must consider the reaction as occurring in two stages. These two stages are shown diagramatically in Figure 5-7. In the first stage all the bonds in the reactant molecules are broken until only atoms are left. The internal energy change for this stage will be positive, since atoms have more potential energy than the molecules from which they are derived (see Figure 3-7). Thus ΔE for the first stage is equal to the sum of all the bond energies of the bonds in the reactant molecules; i.e.,

$$\Delta E_I = \Sigma D_r$$

The second stage corresponds to the formation of the product molecules from the atoms produced in the first stage. Such a formation of bonds corresponds to a decrease in internal energy. Thus ΔE for the second stage is given by the *negative* of the sum of all the bond energies of the products; i.e.,

$$\Delta E_{II} = -\Sigma D_p$$

Adding the internal energy change, we obtain the net value of ΔE for the total reaction:

$$\Delta E = \Delta E_I + \Delta E_{II}$$

(5.18)

$$= \Sigma D_r - \Sigma D_p$$

Equation (5.18) gives us a real insight into why a reaction is exothermic or endothermic. If the sum of the energies of the bonds that are broken is smaller than the sum of the energies of the bonds that are made, then the reaction will

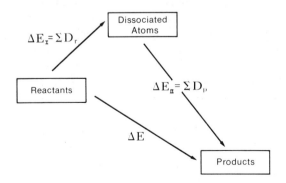

Figure 5-7. A gaseous reaction considered as a two-stage process. In the first stage, all bonds are broken; in the second stage, new bonds are formed.

be *exothermic*. On the other hand, if the reverse is true, the reaction will be endothermic. From a somewhat different point of view, whether or not a reaction is exothermic depends on two factors: (i) the relative *energies* of the bonds broken and made, and (ii) the relative *number* of bonds broken and made. An exothermic reaction corresponds to the formation of *more* bonds, or *stronger* bonds, or both.

Looking at the situation in these terms, it is easy to see why reaction (5.17) is exothermic. The same number of bonds is broken as is made in this reaction, but the bonds that are *made,* C–F and H–F bonds, are conspicuously *stronger* than the C–H and F–F bonds that they replace. A contrary example is provided by the well known reaction

$$2\,H_2(g) + O_2(g) \rightarrow 2\,H_2O(g).$$

It is not the relative *strength* but the relative *number* of bonds made and broken in this reaction that makes it exothermic. The three bonds involved are of very similar energy:

Bond	Bond Energy (kcal mole^{-1})
H–H	104
O=O	119
H–O	111

Nevertheless, the reaction is highly exothermic. The reason for this is simple enough. Only *three* bonds (two H–H bonds and one O–O bond) are broken, while *four* (all O–H bonds) are made. ΔE for this reaction is -115.6 kcal, not very different in value from any of the bond energies involved.

AN IMPORTANT BOND ENERGY

The last example of the use of bond energy tables that we shall consider in this chapter draws attention to a very important bond energy, that of the C–O bond in CO_2. The value of this bond energy is 192 kcal. Among commonly encountered bonds, only the triple bonds such as N≡N, C≡C, and C≡O have larger bond energies. Now, we have just seen above that an endothermic gaseous reaction corresponds among other things to the formation of *stronger* bonds. We should thus expect most of the reactions in which CO_2 is produced to be *exothermic*. This expectation is realized in practice. Very few reactions in which CO_2 is formed are endothermic.

A typical example of a reaction in which CO_2 is a product is the reaction

$$\text{n-}C_7H_{16}(g) + 11\,O_2(g) \rightarrow 7\,CO_2(g) + 8\,H_2O(g) \quad (25°C, 1\ atm) \qquad \textbf{(5.19)}$$

n-Heptane $(n\text{-}C_7H_{16})$ has the structural formula

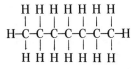

and has 6 C—C bonds and 16 C—H bonds. The occurrence of reaction (5.19) thus corresponds to the breaking of 6 C–C bonds, 16 C–H bonds, and 11 O–O bonds, and the formation of 14 C–O bonds and 16 H–O bonds. The energies involved in these processes are tabulated and added in Table 5-4.

Table 5-4

Bonds and energies involved in the oxidation of n–heptane

No.	Bond	Energy (kcal)	Total Energy	No.	Bond	Energy (kcal)	Total Energy
	Bonds Broken				*Bonds Made*		
6	C–C	83	498	16	O–H	111	1776
16	C–H	99	1584	14	C–O	192	2688
11	O–O	119	1309				
			3391 kcal				4464 kcal

From Table 5-4 we find that ΔE for reaction (5.19) is

$$\Delta E = 3391 - 4464 = -1073 \text{ kcal}$$

(experimentally, $\Delta E = -1077$ kcal). Even though a few more bonds are broken than are made in this reaction (33 versus 30), the fact that the C–O bond in CO_2 is so much stronger than any of the three which are broken is enough to make the reaction highly exothermic.

Reaction (5.19) was not chosen at random. n-Heptane is a constituent of gasoline, and its oxidation to CO_2 and H_2O is typical of that of the other hydrocarbons found in gasoline. If such oxidations were not exothermic, the internal combustion engine could not function. The basic reason for such reactions being exothermic is that they result in the formation of the very strong C–O bonds of CO_2 at the expense of breaking weaker bonds. In fact, the large value of the C–O bond energy in CO_2 has even wider implications than the internal combustion engine. Almost all of the energy we use on this planet today derives from the oxidation of carbon compounds to carbon dioxide. Again, the main reason that the oxidation of wood, oil, coal, natural gas, and other fuels is exothermic is that strong carbon-oxygen bonds are produced.

STRONG AND WEAK BONDS

The case just considered shows how useful it can be to know whether a bond is strong or weak, and raises the question of whether we can easily predict the strength of bonds. Unfortunately, the calculation of bond energies from quantum mechanical principles is a lengthy and not altogether successful process. Even with the aid of large computers, the number of bond energies that can be calculated accurately remains very small. Despite these difficulties, there are several empirical generalizations we can make about the strength of bonds from an inspection of Table 5-3.

The first and most obvious of these generalizations is that multiple bonds are stronger than single bonds. The largest bond energies in Table 5-3 are all associated with triple bonds. As can be seen from the table, triple bonds all have energies of 200 kcal mole^{-1} or more. Next in energy are the double bonds, which range in value from 200 kcal mole^{-1} down to 100 kcal mole^{-1}. Lowest in energy value are the single bonds, most of which have bond energies less than 100 kcal mole^{-1}. The five single bonds that have bond energies in excess of 100 kcal mole^{-1} are rather interesting. They are, in decreasing order, H—F, C—F, H—O, H—H, and H—Cl. All five involve either H or F atoms.

The second generalization we can make about the strength of bonds involves their polarity. Other things being equal, the more polar a bond is the stronger it is. The most obvious examples of this rule are bonds involving the halogens. We notice that the strongest of the hydrogen-halogen bonds is the H—F bond, while the weakest is the H—I bond. Similarly, the carbon-halogen bonds decrease in bond energy value in the order C—F, C—Cl, C—Br, and C—I. This rule about polarity is not without its exceptions, as a close inspection of Table 5-3 will reveal. In particular, it applies rather poorly to multiple bonds.

A very obvious feature of Table 5-3 is the strength of all bonds involving hydrogen. No one such bond has an energy less than 70 kcal mole^{-1}. This cannot be said for any other element. Part of the reason for this behavior is that hydrogen is the most electropositive element featured in the table, but a more important reason is the small size of the hydrogen atom. Even the least polar of the bonds involving hydrogen, namely the H—H bond, is still a strong bond.

It is not always true, however, that smaller atoms give stronger bonds. In particular, the Cl—Cl bond (58 kcal mole^{-1}) is *stronger,* not weaker, than the F—F bond (37 kcal mole^{-1}). In fact, three of the weakest bonds in the whole table involve small atoms. These are the N—N, O—O, and F—F bonds, all of which are between 30 and 40 kcal mole^{-1}. The occurrence of such weak single bonds has a large influence on the chemistry of the three elements nitrogen, oxygen, and fluorine. We shall explore some instances of this in a later chapter.

PROBLEMS 5

1. If ΔE for a reaction is positive, is the reaction exothermic or endothermic?

2. Suggest four intensive properties that a system can have, but which are not mentioned in the text.

3. At $250°K$, the populations of the various levels of the system shown in Figure 5-1 are as follows.

Level	Population (moles)	Energy (kcal mole^{-1})
0	0.634	0.0
1	0.232	0.5
2	0.085	1.0
3	0.031	1.5
4	0.011	2.0
5	0.004	2.5
6	0.002	3.0
7	0.001	3.5

Calculate the value of E for the system at this temperature.

4. Use the results of the previous problem, together with values of E for other temperatures given in the text, to find the average molar heat capacity of the system over the ranges $100-150°K$, $150-200°K$, and $200-250°K$. Is the molar heat capacity constant with temperature? (The molar heat capacity is the energy required to increase the temperature of a mole of substance by $1°C$.)

5. The values of ΔE at $25°C$ and one atmosphere pressure for the stepwise dissociation of carbon tetrafluoride, CF_4, are as follows:

$$CF_4 \; (g) \rightarrow CF_3 \; (g) + F \; (g) \qquad \Delta E = 125.4 \text{ kcal}$$
$$CF_3 \; (g) \rightarrow CF_2 \; (g) + F \; (g) \qquad \Delta E = 94.9 \text{ kcal}$$
$$CF_2 \; (g) \rightarrow CF \; (g) + F \; (g) \qquad \Delta E = 106.8 \text{ kcal}$$
$$CF \; (g) \rightarrow C \; (g) + F \; (g) \qquad \Delta E = 140.0 \text{ kcal}$$

Calculate a mean bond energy for the C—F bond.

6. Use the bond energy values given in Table 5-3 to estimate ΔE for the following reactions:

a. $H_2(g) + I_2(g) \rightarrow 2HI(g)$
b. $4\,HCl(g) + O_2(g) \rightarrow 2Cl_2(g) + 2H_2O(g)$
c. $CH_4(g) + 2O_2(g) \rightarrow CO_2(g) + 2H_2O(g)$
d. $2CO(g) + O_2(g) \rightarrow 2CO_2(g)$
e. $N_2(g) + 3F_2(g) \rightarrow 2NF_3(g)$
f. $H_2O_2(g) \rightarrow H_2(g) + O_2(g)$
g. $2H_2O_2(g) \rightarrow 2H_2O(g) + O_2(g)$
h. $H_2N-NH_2(g) + O_2(g) \rightarrow N_2(g) + 2H_2O(g)$
i. $2\,NF_3(g) \rightarrow F_2N-NF_2(g) + F_2(g)$
j. $CO(g) + H_2O(g) \rightarrow CO_2(g) + H_2(g)$

7. Using a table of bond energy values, estimate ΔE for the reaction

$$CH_4(g) + X_2(g) \rightarrow CH_3X(g) + HX(g)$$

for X= F, Cl, Br, I.

8. In Figure 4–7 (p. 85) five different energy level situations are shown for an equilibrium of the type

$$A(g) \rightleftharpoons B(g)$$

In all five cases, the energy levels of both species are evenly spaced. The interval between levels is either 1 cal mole^{-1} or 2 cal mole^{-1}. The difference in energy between ground states is either zero or 6 cal mole^{-1}.

Match each situation given in the figure with the five following possibilities for the behavior of ΔE.

a) $\Delta E = 0$ at all temperatures.
b) $\Delta E = +6$ cal at all temperatures.
c) $\Delta E = 0$ at $0^\circ K$ but rises to $\Delta E = 0.5$ cal at very high temperatures.
d) $\Delta E = 6$ cal at $0^\circ K$ but rises to $\Delta E = 6.5$ cal at very high temperatures.
e) $\Delta E = 6$ cal at $0^\circ K$ but falls to $\Delta E = 5.5$ cal at very high temperatures.

9. From the following data, all of which can be obtained directly from experiment, find ΔE for the reaction

$$C\,(diamond) \rightarrow C(g) \quad (25^\circ C,\ 1\ atm)$$

by the use of Hess's Law.

Reaction	ΔE (for 25°C and 1 atm) (kcal)
$C(diamond) + O_2(g) \rightarrow CO_2(g)$	-94.5
$CO(g) + \frac{1}{2}O_2(g) \rightarrow CO_2(g)$	-67.3
$O_2(g) \rightarrow 2O(g)$	$+118.5$
$CO(g) \rightarrow C(g) + O(g)$	$+256.7$

10. Why is the value of ΔE obtained in the previous problem very close to twice the bond energy of a carbon-carbon single bond?

11. When one mole of Zn metal is added to 2M HCl solution at $25°C$ and 1 atmosphere pressure, the metal dissolves and hydrogen gas is evolved. At the same time, 36.4 kcal of heat energy are released.
  a) Calculate ΔV for the reaction

$$Zn(s) + 2H^+(2M) \rightarrow Zn^{2+}(1M) + H_2(g)$$

(Ignore the difference in volume between 2M HCl and 1M $ZnCl_2$.)
  b) Calculate the expansion work $P\Delta V$ in both liter-atmospheres and calories.
  c) Calculate ΔE for the above reaction.

12. The volume of 1 g of liquid mercury is 0.07355 ml at $0°C$ and 0.07489 at $100°C$. The molar heat capacity of mercury over this range is 6.60 cal deg^{-1} $mole^{-1}$. Use these data to calculate $w = P\Delta V$, q, and ΔE for the process

$$Hg\ (0°C,\ 1\ atm) \rightarrow Hg\ (100°C,\ 1\ atm).$$

13. ΔE for the reaction

$$3\ O_2(g) \rightarrow 2\ O_3(g) \qquad (25°C,\ 1\ atm)$$

has the value +34.7 kcal. Use this information and the bond energy of O_2 to find the mean bond energy of the O—O bond in ozone. Suggest why this value is intermediate between the bond energy of a single and a double oxygen-oxygen bond.

Chapter 6

THE
ENTROPY

INTRODUCTION

In Chapter 4 we learned to look at chemical equilibrium as a competition between alternative sets of energy levels, and we saw that the set with the *lower-lying* levels was favored at *low* temperatures while the set with the more *closely-spaced* levels was favored at *high* temperatures. We also saw that, for all the insight such an approach gave us into chemical equilibrium, it has one serious shortcoming: it is difficult to apply to real cases of chemical equilibrium. The complexity of handling all the energy levels involved is too great.

A way out of this difficulty was then suggested: to try to relate our microscopic view of equilibrium to the *bulk behavior* of matter. Once such a relationship is established, we no longer have to think in detail about all the energy levels involved.

We are now half way through such a process of changing from a microscopic to a macroscopic way of dealing with chemical equilibrium. In the previous chapter we became acquainted with a quantity E, called the internal energy. The internal energy is related to energy levels and populations through the equation

$$E = \Sigma \epsilon_i n_i$$

on the one hand, and is related to the bulk behavior of matter through the equation

$$\Delta E = q - w$$

on the other. Moreover, E is one of the factors which determine the position of chemical equilibrium. That side of a chemical reaction having the lower energy levels is also that side with the lower internal energy. We can thus expect that side of a chemical equilibrium with the lower value of E to be favored at low temperatures. In other words, if ΔE for a reaction is *positive,* we expect the *reactants* to be favored at low temperatures, while if ΔE is *negative* we expect the *products* to be favored at low temperatures.

What we have done for the *relative energies* of the two competing sets of energy levels we must now do for the *relative spacings*. We must find another quantity which behaves rather like E and ΔE, except that it tells us about *spacings* instead of about the energies of sets of levels. We will then be able to predict the behavior of chemical equilibria at high temperatures. In this chapter we shall introduce such a quantity. It is called the *entropy* and is given the symbol S.

W_{max}

Before dealing with entropy directly, we must first introduce another quantity closely related to it. This is the quantity W_{max}.

In Chapter 2 we discussed how particles can be arranged among energy levels in various distributions, all of which are consistent with the same total energy. We noted that there is a number W, associated with each distribution, called its thermodynamic probability. W gives the number of microstates corresponding to each distribution, and hence is proportional to the probability of each distribution.

The most probable of all these distributions is the so-called Boltzmann distribution. The form of this distribution is given by the Boltzmann Law, equation (2.5). Because the Boltzmann distribution is the most probable distribution, the value of W associated with it is larger than for any other distribution. Accordingly, this value of W is given the special symbol W_{max}, the subscript being an abbreviation for the word maximum.

This quantity W_{max} has several interesting and useful properties, which we will now explore. The first of these properties is its dependence on the number of levels which are occupied. We find that *the greater the number of energy levels occupied, the bigger the value of W_{max}.*

The best way of seeing why W_{max} behaves in this way is by taking a simple example involving only a few particles. The example we shall take is shown in Figure 6-1. There we compare two simple systems, each with the same quantity of energy and each involving six particles. In both systems the energy levels are evenly-spaced. The levels are twice as far apart in one case as in the other. The

System 1   System 2

Figure 6-1

most probable distribution for each system is shown.

Strictly speaking, neither of these distributions is a Boltzmann distribution, since we can only talk about a Boltzmann distribution when a very large number of particles is involved. Nevertheless, they are both the most probable distributions for the given circumstances and sufficiently approximate a Boltzmann distribution for our purposes.

We can calculate the value of the thermodynamic probability for each system in the figure by using equation (2.3).

For system 1,

$$W_1 = \frac{6!}{3!\, 2!\, 1!} = 60$$

while for system 2,

$$W_2 = \frac{6!}{4!\, 2!} = 15$$

W_1 is four times as large as W_2. As predicted, the system with the larger number of occupied levels has the larger value of W.

The reason why W_1 is larger than W_2 is now fairly obvious. Both the above expressions for W_1 and W_2 have the same numerator, namely 6!. They differ in the value of the denominator. With three levels occupied, the denominator has the value 3! × 2! × 1! = 12, while with only two occupied it has the value 4! × 2! = 48. Such behavior is general for Boltzmann distributions. The larger the number of levels occupied, the smaller will be the value of the denominator in the expression

$$W_{max} = \frac{n!}{n_0!\, n_1!\, n_2!\, n_3! \dots}$$

Thus, if we compare two Boltzmann distributions, each with the same number of particles, and in one distribution more levels are occupied than in the other, the distribution with the larger number of occupied levels is invariably the one with the larger value of W_{max}.

An immediate corollary of this rule is that W_{max} *for a system must rise with temperature.* As the temperature is raised, levels which were previously unoccupied become occupied; i.e., *more* levels become occupied. An example of such behavior is shown in Figure 6-2. The distribution of one mole of HCl molecules over vibrational levels is illustrated both at 25°C and at 300°C. The value of W_{max} at the higher temperature is higher than that at the lower temperature for no other reason except that more energy levels are occupied at the higher temperature.

A second corollary, more central to our present purposes, also follows from

Level

Level	Populations at 25°C	Populations at 300°C
8		57
7		1.32×10^4
6		3.92×10^6
5		1.51×10^9
4	8	7.48×10^{11}
3	1.95×10^6	4.81×10^{14}
2	8.02×10^{11}	4.00×10^{17}
1	5.41×10^{17}	4.31×10^{20}
0	6.02×10^{23}	6.02×10^{23}

Figure 6-2. Distribution of a mole of HCl molecules over vibrational energy levels at 25°C and at 300°C.

$$W_{max} = 10^{(3 \times 10^{18})} \qquad W_{max} = 10^{(3 \times 10^{21})}$$

our rule: *If two systems with the same number of particles at the same temperature are compared, that system with the more closely spaced energy levels will have the larger value of W_{max}.*

In Figure 6-3 we illustrate an example of such behavior. The vibrational distributions of one mole of HCl gas and of one mole of HI gas for 25°C are compared. Higher, unoccupied levels are not shown. As the numbers at the bottom of the figure indicate, W_{max} for one mole of HI is larger than W_{max} for one mole of HCl. The vibrational levels of HI are closer to each other than are the vibrational levels of HCl. As a result there are more HI levels occupied at 25°C than there are HCl levels occupied. Since *more* levels are occupied, W_{max} is *larger* for HI.

We can expect this behavior to be quite general. If two systems, each with the same number of particles and each at the same temperature, are compared, that system with the more closely spaced levels will always have the larger number of occupied levels and hence the larger value of W_{max}. Thus W_{max} has just the property we want it to have. It depends on the spacing of the energy levels.

In addition to its dependence on the spacing of levels, W_{max} has another property which is of interest to us. When two independent systems are considered together, the values of W_{max} for each system must be *multiplied* in

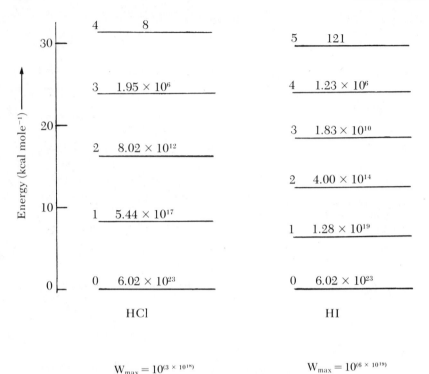

$$W_{max} = 10^{(3 \times 10^{18})}$$

$$W_{max} = 10^{(6 \times 10^{19})}$$

Figure 6-3. Boltzmann distributions of one mole each of HCl and HI molecules at 25°C.

order to yield the value of W_{max} for the combined system.

Given any two independent systems, we can label them a and b. Let W_{max} (a) be the thermodynamic probability of the Boltzmann distribution of System a, and W_{max} (b) the equivalent quantity for System b. Further, let W_{max} (tot) be the thermodynamic probability of both systems considered together, each in its most probable distribution. If the two systems are independent, any of the W_{max} (a) microstates of System a can be combined with equal probability with any of the W_{max} (b) microstates of System b. This gives a total of W_{max} (a) $\times$ W_{max} (b) ways of achieving both systems in their most probable state. In other words:

$$W_{max} \text{ (tot)} = W_{max} \text{ (a)} \times W_{max} \text{ (b)} \tag{6.1}$$

THE ENTROPY

Although W_{max} has several convenient properties, chief among them its dependence on the spacing of energy levels, it is not W_{max} itself, but a quantity proportional to its logarithm which is used in thermodynamics. This second quantity is called the *entropy* and is given the symbol S. The entropy is used in preference to W_{max} for several reasons. One of these is historical. The entropy was discovered to be a useful property long before the quantization of energy was appreciated. Furthermore, S is a quantity which is easier to manipulate than W_{max}. Finally, as we shall presently establish, entropy can be simply related to heat changes, and thus to the bulk properties of a system.

The entropy of a system is defined by the relationship

$$S = k \log_e W_{max} \tag{6.2}$$

where k is the now familiar Boltzmann constant. Notice that the logarithm referred to in equation (6.2) is to the base e; i.e., it is a *natural logarithm*.

In order to familiarize ourselves with equation (6.2), let us calculate a value for S from a given value of W_{max}. Typical values of W_{max} for a mole of a pure substance fall between

$$(10)^{10^{24}} \text{ and } (10)^{10^{25}}$$

Let us use the smaller value,

$$W_{max} = (10)^{10^{24}}$$

Taking logarithms to the base ten we then find that

$$\log_{10} W_{max} = 10^{24}$$

Natural logarithms and base ten logarithms can be interconverted by means of the relationship

$$\log_e x = \log_e 10 \times \log_{10} x = 2.303 \log_{10} x$$

Thus:

$$\log_e W_{max} = 2.303 \log_{10} W_{max} = 2.303 \times 10^{24}$$

Using $k = 3.299 \times 10^{-24}$ cal deg^{-1}, we find that:

$$k \log_e W_{max} = (3.299 \times 10^{-24})(2.303 \times 10^{24}) \text{ cal deg}^{-1}$$

$$= 7.60 \text{ cal deg}^{-1}$$

Thus the entropy of the system is given by

$$S = 7.60 \text{ cal deg}^{-1}$$

If we had used the higher value for W_{max}, namely $(10)^{10^{25}}$, we would have obtained a value of 76.0 cal deg^{-1} for the entropy. A mole of substance typically has an entropy between 0 and 100 cal deg^{-1}. Values of the entropy thus involve much simpler numbers than do values of W_{max}.

Notice that the dimensions of entropy are energy divided by temperature. This follows from the definition, $S = k \log_e W_{max}$. W_{max} is a pure number and so is its logarithm. S thus has the same dimensions as k, namely energy/temperature.

SOME PROPERTIES OF THE ENTROPY

We come now to consider some of the properties of the function just defined. In some ways the entropy behaves very much like the quantity W_{max} but in other ways it behaves rather differently. The major point of similarity between the two quantities is that both *increase under exactly the same circumstances.* This behavior derives from the nature of logarithms. The function $k \log W_{max}$ is what mathematicians call an increasing function of W_{max}; if W_{max} increases so does the function $k \log W_{max}$.

We have already established that for a given number of particles, W_{max} increases as the number of energy levels which are occupied increases. The same behavior therefore must also be true of the entropy. From this we can quickly conclude for S as we have already done for W_{max}, that

i) *The entropy of a given system always rises with temperature, and*

ii) *If two systems with the same number of particles at the same temperature are compared, that system with the more closely spaced energy levels will have the larger value of S.* The entropy of a system thus has the useful characteristic of depending on the spacing of the energy levels of the system.

A third property of the entropy is that S is a *state function.* If the state of a system is defined, then the paricles comprising it are in a unique Boltzmann distribution. As a result, both W_{max} and $k \log_e W_{max}$ have particular values. On the other hand, if the state of a system changes so, in general, will the populations of the various energy levels, and with them the values of W_{max} and S.

The entropy is not only a state function but also an *extensive property.* Consider two systems, System a and System b. Let S_a be the entropy of System a and S_b the entropy of System b. Further, let S_{tot} be the entropy of the two systems considered together. We then have:

$$S_a = k \log_e W_{max}(a)$$

$$(6.3)$$

$$S_b = k \log_e W_{max}(b)$$

while

$$S_{tot} = k \log_e W_{max}(tot)$$

Now, we know from equation (6.1) that

$$W_{max}(tot) = W_{max}(a) \times W_{max}(b)$$

Taking logarithms we then obtain

$$\log W_{max}(tot) = \log W_{max}(a) + \log W_{max}(b)$$

irrespective of the base of the logarithms. Multiplying through by k then gives us:

$$k \log_e W_{max}(tot) = k \log_e W_{max}(a) + k \log_e W_{max}(b) \qquad \textbf{(6.4)}$$

Substituting equations (6.3) into equation (6.4) yields the final result:

$$S_{tot} = S_a + S_b \qquad \textbf{(6.5)}$$

Equation (6.5) tells us that the entropy of two systems considered together is equal to the sum of the entropies of the individual systems. This must mean that the entropy of *two moles* of a substance at a given temperature and pressure is equal to *twice* the entropy of *one mole* of the substance at the same temperature and pressure. Taking such an argument one stage further, it follows that the entropy of n moles of substance is equal to n times the entropy of one mole of substance under the same conditions. In other words, entropy is an *extensive property*.

Because the entropy, like the internal energy, is both a state function and an extensive property, we can add and subtract entropy values in the same way as we can add and subtract values of the internal energy in accord with Hess's Law. In order to do this we need to have access to some values of the entropy for different substances. Table 6-1 provides such data. A more comprehensive table can be found in Appendix 3 at the end of the book.

The entropy values quoted in Table 6-1 are what are known as *standard molar entropies*. The standard molar entropy of a substance is the entropy of one mole of that substance at one atmosphere pressure. Such a quantity is given the symbol S^0, the superscript indicating the *standard* condition of one atmosphere pressure. Since the entropy is a state function, not only the pressure but also the temperature must be specified. Entropy values given in Table 6-1 are for 25°C. The temperature for which a standard molar entropy is given is sometimes indicated by a subscript. For example, S^0 values for 25°C (298°K) are often indicated as S^0_{298}.

Table 6-1

The Standard Molar Entropies of Various Substances
(in cal deg^{-1} mole^{-1} at 25°C and 1 atm)

Solids		Gases	
C (diamond)	0.58	He	29.8
Si	4.5	H_2	31.2
LiF	8.6	D_2	34.6
Ge	10.1	Ne	34.9
Sn (gray)	10.6	Ar	36.7
LiCl	13.2	Xe	40.5
NaCl	17.3	HCl	44.6
KCl	19.8	H_2O	45.1
KI	24.9	N_2	45.8
I_2	27.9	NH_3	46.0
CsI	31.0	CO	47.3
		F_2	48.4
		O_2	49.0
		HI	49.3
		NO	50.3
		CO_2	51.1
		Cl_2	53.3
		CH_3CH_3	54.9
		NO_2	57.5
		Br_2	58.6
Liquids		I_2	62.3
		C_5H_{10}	70.0
H_2O	16.7	(cyclopentane)	
Hg	18.5	N_2O_4	72.7
Br_2	36.4	C_5H_{10}	83.1
CCl_4	51.3	(1–pentene)	

Let us use Table 6-1 to calculate the entropy change that accompanies a reaction. Consider, for instance, the reaction

$$H_2(g) + I_2(s) \rightarrow 2HI(g) \quad (25°C, 1 \text{ atm}) \qquad \textbf{(6.6)}$$

From Table 6-1 we quickly find that the entropy of the product (at 25°C and 1 atmosphere) is $2 \times 49.3 = 98.6$ cal deg^{-1}, while that of the reactants at the same temperature and pressure is $31.2 + 27.9 = 69.1$ cal deg^{-1}. The change in entropy is thus given by

$$\Delta S^0 = 98.6 - 69.1 = 29.5 \text{ cal deg}^{-1}$$

Values of ΔS^0 for a whole variety of reactions can be obtained in a similar way* from a table of S^0 values such as is given in Table 6-1 and Appendix 3. The

---

*There are some difficulties about manipulating the entropy which are not found in the case of the internal energy. We will consider these in a later chapter.

value of ΔS^0 for a reaction, obtained in this manner, is a very useful quantity. As we will see later on, it enables us to predict whether or not a reaction will occur at high temperatures.

SOME COMPARISONS OF STANDARD
MOLAR ENTROPY VALUES

The entropy values listed in Table 6-1 all refer to a mole of atoms or molecules at 25°C. Now, we have seen that for a given number of particles at a fixed temperature, the entropy increases the more closely spaced the energy levels are. We therefore expect that the differences in the values of S^0 given in Table 6-1 reflect differences in *the spacing of the energy levels* of the different substances.

We have already discussed in some detail in Chapter 3 what factors govern the spacing of energy levels. First, there is the mass. Other things being equal, the larger the mass of a particle, the more *closely spaced* will be the energy levels which it can occupy. Second, there is the constraint. The larger the forces acting on a body confining it to a smaller region of space, the more *widely spaced* will be the energy levels which it can occupy. An understanding of these two factors enables us to explain many features of Table 6-1. Given two similar compounds, we can usually tell which of the pair will have the more closely spaced levels. Such information will also tell us which will have the higher standard molar entropy. Let us now go through Table 6-1 comparing entropies in this way. Such a process will give us more of an intuitive "feel" for entropy and make it easier for us to use this property in Chapter 7, where we will return to a discussion of chemical equilibrium.

We will begin by comparing the standard molar entropies of He (29.8 cal deg^{-1}) and Ar (36.7 cal deg^{-1}). Both of these gases are monatomic. This means that we need only consider translational levels in comparing the two entropies. Since both gases are at 25°C and one atmosphere, they occupy the same volume. Thus the constraint on the translational motion of the atoms is the same for both gases. However, the spacing of the translational levels in the two cases is different because of a difference in *mass*. Since argon atoms are heavier than helium atoms, we expect them to occupy more closely-spaced levels. This leads us to predict, correctly, that Ar has a larger standard molar entropy than He.

He (29.8 cal deg^{-1}) and D_2 (34.6 cal deg^{-1}) also make an interesting comparison. The difference in entropy here cannot be due to the spacing of translational levels, since the two molecules have virtually identical masses. D_2 has a bigger entropy because its molecules can have rotational energy as well as translational energy. This extra set of energy levels which D_2 molecules can occupy results in a rotational contribution to the entropy of D_2. Such a contribution is lacking in the case of He. (It is also true that D_2 can vibrate, but at 25°C the vibrational contribution is small (1.5×10^{-5} cal deg^{-1}) because these energy levels are so widely spaced that, as we saw in Chapter 3, very few molecules are vibrationally excited.)

Looking next at the gases listed in Table 6-1 as a whole, there are two features which we should observe. The first of these is the general increase in entropy with molecular weight for comparable molecules. We notice, for instance, that among the diatomic molecules the heaviest, such as I_2 and Br_2, have entropies which are almost twice as big as the lightest, namely H_2. The reason for this is obvious. The larger the mass of the molecule, the more closely spaced its energy levels, particularly its translational levels, will be.

The second feature we should notice in Table 6-1 is the increase in entropy with increasing molecular complexity. In particular, the entropies of diatomic gases are almost all larger than those of monatomic gases. The only exceptions are the two lightest diatomic molecules, H_2 and D_2. The reason for this behavior has already been discussed in comparing the entropies of D_2 and He. While the entropies of monatomic gases are purely translational in origin, those of diatomic gases have rotational and to a lesser extent vibrational contributions as well.

It is also obvious from the table that polyatomic gases have higher entropies on the whole than diatomic gases. Partly this is because such molecules are heavier and bigger, giving more closely spaced translational and rotational levels, but there is also another reason. The *vibrational* contribution to the entropy of a polyatomic gas is often significant.

The vibrations of a polyatomic molecule are more complex than those of a diatomic molecule. While vibrations in the diatomic case only involve changes in internuclear distances, those in polyatomic molecules can involve changes in angles as well. Some typical examples of the vibrations of polyatomic molecules are shown in Figure 6-4. The energies quoted there refer to the interval between the two lowest levels in each case.

A vibration of a polyatomic molecule may be defined as a periodic movement which results in a change of size or shape. This is in contrast to a rotation, which causes no change in size or shape. In general, the more atoms a molecule has, the more ways there are in which it can vibrate. There is a simple equation connecting the two. If a molecule has n atoms, then it can vibrate in $3n - 6$ different ways. The water molecule, for instance, has 3 atoms. It therefore has $3 \times 3 - 6 = 3$ different modes of vibration. These three are labeled I, II, and III in Figure 6-4.

Modes of vibration which correspond to alterations in bond lengths are usually called *stretching* vibrations. Examples of such vibrations are numbers I, II, and IV in the figure. Note that these stretching vibrations have more widely spaced levels than the other three vibrations shown. This is usually the case in polyatomic molecules, though there are many exceptions.

The vibration with the most closely spaced levels in the figure is number VI. This vibration corresponds to the internal rotation of one CH_3 group with respect to another around a carbon-carbon single bond in a molecule of ethane.

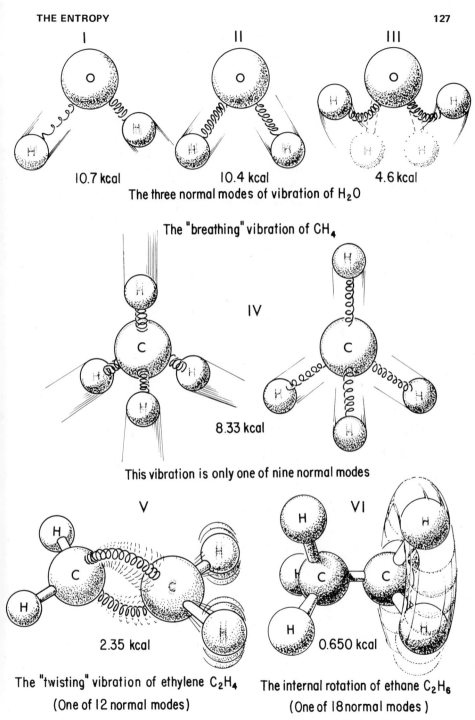

I

10.7 kcal

II

10.4 kcal

III

4.6 kcal

The three normal modes of vibration of H_2O

The "breathing" vibration of CH_4

IV

8.33 kcal

This vibration is only one of nine normal modes

V

2.35 kcal

The "twisting" vibration of ethylene C_2H_4
(One of 12 normal modes)

VI

0.650 kcal

The internal rotation of ethane C_2H_6
(One of 18 normal modes)

Figure 6-4. Typical modes of vibration of polyatomic molecules. The energy quoted for each vibration is the spacing (per mole) between the two lowest levels.

Although such a movement has many of the characteristics of a rotation, it is technically a vibration, since it changes the shape of the molecule. Whatever we call such a movement, though, there can be very little constraint on it. The energy levels are nearly as closely spaced as the rotational levels of small molecules.

The small constraint on rotation about carbon-carbon single bonds, not only in ethane but in more complicated molecules, has very wide repercussions in chemistry. In this book, however, we are interested solely in its effect on the value of the entropy of a substance. A nice instance of this is found if we compare the entropies of two hydrocarbons listed in Table 6-1 whose molecules are illustrated in Figure 6-5. Compound I is the substance cyclopentane $C_5 H_{10}$, containing a ring of five carbon atoms. The rigidity of this ring prevents any rotation around the carbon-carbon single bonds. Compound II is an isomer called 1-pentene. Here there are three carbon-carbon single bonds and no ring preventing the rotation of various segments of the molecule around these three bonds. Note, however, that the double bond is rigid and does not allow rotation of the end carbon.

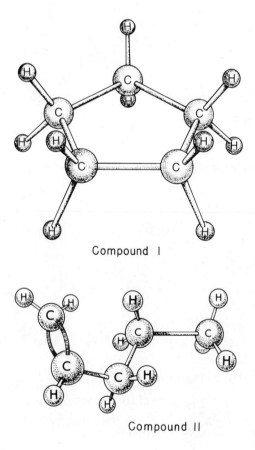

Figure 6-5. Molecular configurations of cyclopentane *(top)* and 1-pentene *(bottom)*.

Compound I

Compound II

The molecular weights of the two isomers are identical, so there will be no difference in the translational contribution to the entropy. Since the sizes of the two molecules are not very different, the difference in rotational entropy between the two isomers will be small. Nevertheless there is a substantial difference in entropy between the two compounds. From Table 6-1 we find that compound 1 has an S^0 value of 70.0 cal deg^{-1}, while compound II has an S^0 value of 83.1 cal deg^{-1}. The higher value for compound II must be due almost entirely to its lower vibrational constraint and the resulting larger vibrational entropy.

We can now turn from a discussion of the entropies of gases to the entropies of solids. The most casual glance at Table 6-1 is enough to reveal that the standard molar entropies of solids are much smaller than those of gases. This is not altogether surprising. Atoms and molecules in solids have little or no translational motion and it is the *translational* part of the entropy which is the major contributor to the entropies of gases.

As our first example of a solid we will consider the element carbon in the form of diamond. In a diamond crystal each carbon atom is held in place by *four* covalent bonds connecting it to four other carbon atoms. To a first approximation we can consider the carbon atoms to vibrate independently of each other about a mean position. These vibrations are very "stiff" because of the large number of covalent bonds confining the atom to a small volume in the crystal lattice. The vibrational levels are thus widely spaced. In consequence the entropy is very low, the lowest of any substance at 25°C.

Notice that the Group IV elements listed in Table 6-1 (Se, Ge, Sn), all of which crystallize in a diamond-type lattice*, have larger entropies than diamond, and that this entropy increases with atomic weight. This increase partially reflects the increase in the mass of the vibrating atoms, but it is also a result of the decrease in the strength of the bonds between the atoms and hence in the constraint on their vibration.

A similar effect can be noticed among the alkali halides. LiF, with small ions tightly held to each other by a strong electrostatic attraction, has the lowest entropy. By contrast, CsI has the largest entropy of the alkali halides, partly because the ions are the heaviest but also because they are the largest and the most loosely held by the attraction of their unlike charges.

To end our discussion of Table 6-1, let us turn finally to consider the standard molar entropy values for liquids. We find that, on the whole, the liquids shown in Table 6-1 have higher entropies than the solids. The reason for this is that the constraint on molecular motion is not as large in liquids as it is in solids. Similarly, liquids usually have smaller molar entropies than gases, though it is noticeable from the table that liquids with very heavy and complex molecules can have higher standard molar entropies than some simple gases. Note that

---

*The entropy value given in Table 6-1 is for gray tin, which is unstable at 25°C but has a diamond-type lattice.

water has a particularly low standard molar entropy, presumably because of the large constraint on molecular motion caused by hydrogen bonding.

In discussing the entropy values in Table 6-1, we have often separated entropies into translational, rotational, and vibrational contributions, without justifying this procedure. The justification of such a division depends ultimately on the independence of the probability of the distribution of particles among different kinds of energy levels. If a molecule has a particular translational energy, this has no influence on its rotational or vibrational energy and vice versa. The different kinds of energy and the distributions corresponding to them are *independent of each other.* We can thus handle the situation in the same way that we handled independent systems earlier in the chapter. Corresponding to equation (6.5), which we derived there,

$$S_{tot} = S_a + S_b$$

we can now write

$$S_{tot} = S_{trans} + S_{rot} + S_{vib} \qquad \text{(6.7)}$$

In both cases we have different distributions which are independent of each other. This allows us to *add* their contributions to the total entropy.

Table 6-2 gives some examples of the magnitudes of the translational, rotational, and vibrational contributions to the standard molar entropy of some diatomic gases at 25°C. Notice that the translational contribution is always the major contributor to the entropy. For gases with large and heavy molecules, the rotational contribution can be quite large (29 per cent in the case of I_2). The vibrational contribution, by contrast, is always small and often negligible.

Table 6-2

The standard molar entropies of some diatomic gases at 25°C
(cal deg^{-1} mole^{-1})

Gas	Translational Contribution	Rotational Contribution	Vibrational Contribution	Total Entropy
H_2	28.10	3.10	0.00	31.20
N_2	35.96	9.83	0.00	45.79
Cl_2	38.72	14.02	0.39	53.13
I_2	42.53	17.75	2.00	62.28
HCl	36.73	7.87	0.00	44.60
HI	40.47	8.82	0.00	49.29

THE SIGN OF ΔS^0 FOR A REACTION

The sign and magnitude of ΔS^0 for a given reaction are useful things to know. As has already be remarked, the sign of ΔS^0 helps us to predict the direction in which a reaction will go. An accurate value of ΔS^0 can always be obtained from a sufficiently comprehensive table of standard molar entropies. For approximate purposes, though, the exact magnitude of ΔS^0 is not always required. It is often sufficient to know whether ΔS^0 for a reaction is a positive or negative quantity. In a surprising number of cases, we can tell this without the necessity of consulting tables. We need only draw on the insight we have just acquired into the factors governing the value of the standard molar entropy of a substance.

In general, a reaction will be accompanied by an increase in entropy if it corresponds to the *relaxation of some of the constraints* on the particles involved in the reaction. (The mass, of course, is unaltered in a chemical reaction.) Conversely, a reaction will result in a decrease in the entropy if the products are more "constrained" than the reactants.

The most obvious example of a type of chemical reaction in which the entropy increases is a *dissociation reaction.* We have already discussed such reactions in Chapter 4 and seen that dissociation corresponds to the relaxation of a constraint and a movement from widely spaced to closely spaced energy levels. Such a reaction is thus one in which the entropy increases.

There is another way of rationalizing the fact that the entropy increases in a dissociation reaction. Take, for instance, the case

$$N_2O_4(g) \rightarrow NO_2(g) + NO_2(g) \qquad (6.8)$$

Here a molecule is broken up into smaller fragments and the *number of moles increases* as a result of the reaction occurring. We have seen in Table 6-2 that in the case of gases the translational contribution is the largest contribution to a standard molar entropy. It is also apparent from the table that although the translational entropy increases with the mass of the molecule, it does so rather slowly. For instance, the translational entropy of I_2 gas is only about 1-1/2 times that of H_2, despite the fact that I_2 is 127 times as heavy as H_2. If we now make the rather gross approximations that (a) non-translational contributions to the entropy are negligible and (b) the translational entropy is almost independent of mass, we arrive at the conclusion that *if the number of moles of gas increases in a gaseous reaction, so does the entropy.* Despite the character of the approximations involved, such a statement is a valuable rule of thumb.

The application of this rule is not limited to dissociation reactions. For instance, the reaction

$$C_5H_{12}(g) + 8\,O_2(g) \rightarrow 5\,CO_2(g) + 6\,H_2O(g) \quad (25^\circ C, 1\,atm)$$

is not a dissociation reaction. It is, however, a reaction in which the number of moles of gas increases (from 9 to 11). We should thus expect ΔS^0 to be positive. Using the values listed in Appendix 3, we indeed find that ΔS^0 is positive, about +60.8 cal deg^{-1}.

Since the constraint on particles is less in gases than it is in liquids, and less in liquids than in solids, we expect to find that entropies of vaporization and fusion are positive. For example, consider the vaporization of water:

$$H_2O(1) \rightarrow H_2O(g)$$

for which ΔS^0 is +28.4 cal deg^{-1} at 25°C. By a similar argument, we can predict an entropy increase when gases are formed at the expense of liquids or solids, or when liquids are formed at the expense of solids. Thus, for instance, in the reaction

$$H_2(g) + I_2(s) \rightarrow 2\,HI(g) \tag{6.9}$$

since two moles of gas have been formed from one mole of gas plus one of solid we properly predict that there is an entropy increase.

Although the approach described above enables us to predict the sign of ΔS^0 in a large number of cases, there still remain many reactions in which ΔS^0 is rather small and impossible to predict without tables. A case in point is the reaction

$$CO(g) + NO_2(g) \rightarrow CO_2(g) + NO(g)\quad (25°C, 1\ atm) \tag{6.10}$$

Since the number of moles is the same on both sides of this reaction, we cannot easily decide whether the entropy change is positive or negative. We can, however, predict that it will be small. From Table 6-1, $\Delta S^0 = -3.4$ cal deg^{-1}.

MEASURING THE ENTROPY

Now that we have attained some familiarity with the entropy, have some feeling for its magnitude, and can predict whether it will increase or decrease in a given reaction, we may well ask how values of this property are measured. This is a difficult question to answer. Unfortunately there is no simple, direct way of measuring the entropy of a system analogous to that of measuring its mass or its volume. There is no convenient black box, labeled "Entropy Meter," into which we can put a substance, and read its entropy on a dial. We can only arrive at a value for the entropy by calculations based on a large number of measurements of other properties of a system. In such circumstances it is perhaps best to talk about *evaluating* the entropy of a substance rather than measuring it.

The fact that we cannot measure the entropy easily or directly is unfortunate. Such a state of affairs quite often persuades students that there is something phony about entropy. (Even some chemistry professors share this attitude.) The feeling is that a property which is so inaccessible to direct measurement must be less real than a property such as mass or volume, which even a high school student can measure. Worse still, some people have the idea that the entropy is a "concept" rather than a property of a substance. The beginning student must be warned against acquiring such misconceptions. The entropy of a substance is a property of everyday utility in chemistry. That it is difficult to measure does not detract from this fact.

There are two major methods for evaluating the entropy. The first of these methods leads to what is usually called a *spectroscopic value* for the entropy, since it involves calculating S from energy level data derived from spectroscopy. It is only possible to calculate spectroscopic values for gases. The second method for finding S leads to a *calorimetric value*. In this method energy levels are not considered at all. The value of the entropy is derived from an experimental study of heat changes. Data such as heat capacities over a range of temperatures and heats of fusion and vaporization are required. We will consider each of these two methods in turn.

Spectroscopic Values of the Entropy

Suppose we have a system for which we know the population of each energy level. Let these populations be n_0, n_1, n_2, n_3, ... The knowledge of these populations allows us to calculate W_{max} for the system using formula (2.3),

$$W_{max} = \frac{n!}{n_0! \, n_1! \, n_2! \, n_3! \, \ldots}$$

Once W_{max} has been found, the entropy can then be calculated from relationship (6.2),

$$S = k \, \log_e W_{max}$$

Such a sequence of calculations provides us with a method for calculating the entropy of the system.

The simplest example of such a calculation is the case of a pure crystalline substance at the absolute zero of temperature. Here all the atoms or molecules are in the ground state. No other energy level is occupied. If the crystal consists of a mole of atoms, then the total number of particles is given by the Avogadro number, N. The population of the ground level will also be equal to N, or

$$n_0 = N$$

Substituting into equation (2.3), we obtain:

$$W_{max} = \frac{N!}{N!} = 1$$

from which it follows that

$$S = k \log_e 1 = 0$$

The entropy of the crystal is thus *zero at the absolute zero* of temperature. The assertion that *all* perfectly crystalline substances have zero entropy at $0°K$ is sometimes called the *Third Law of Thermodynamics*.

Another simple example of an entropy calculation with the use of equations (2.3) and (6.2) is the calculation of the *electronic* contribution to the entropy in the case of oxygen gas. O_2 is one of those few molecules whose ground state is *degenerate*. We cannot enter into the precise reasons for this here, except to say that it is connected with the fact that O_2 has two unpaired electrons. These can orient their spins in three different ways, giving rise to *three* levels of almost identical energy.

Since we are concerned only with the electronic contribution to the entropy of O_2, we need not consider other kinds of levels. We need only deal with electronic levels. In effect this means dealing only with the three equivalent ground levels. The excited electronic states of O_2 are too high in energy to be populated at ordinary temperatures.

If the three ground levels are equal in energy, they will have equal populations. For a mole of O_2 molecules, the population of each level will be $N/3$, where N is the Avogadro number. Thus

$$W_{max} = \frac{N!}{(N/3)! \, (N/3)! \, (N/3)!} \tag{6.11}$$

To evaluate these large factorials we make use of *Stirling's approximation*, which states that:

$$\log_e(N!) = N \log_e N - N$$

Taking natural logarithms of equation (6.11) and substituting in Stirling's approximation, we find:

$$
\begin{aligned}
\log_e W &= \log_e(N!) - 3 \log_e \left\{ (N/3)! \right\} \\
&= N \log_e N - N - 3 \left\{ (N/3)\log_e(N/3) - N/3 \right\} \\
&= N \log_e N - N - \left\{ N \log_e(N/3) - N \right\} \\
&= N \log_e N \quad - \quad N \left\{ \log_e N - \log_e 3 \right\} \\
&= N \log_e 3
\end{aligned}
$$

Thus:

$$S_{elec} = k \log_e W = kN \log_e 3 = R \log_e 3$$

$$= 1.987 \times 1.099 \text{ cal deg}^{-1}$$

$$= 2.18 \text{ cal deg}^{-1}$$

It is apparent from Table 6-1 that the entropy of O_2 is larger than can be accounted for purely in terms of translational, rotational, and vibrational energy levels. O_2 has a mass, an internuclear distance, and a bond energy intermediate between those of N_2 and F_2. We should thus expect the translational, the rotational, and the vibrational entropies also to be intermediate. Yet the S^0 value of O_2 is 49.0 cal deg^{-1}, which is *not* intermediate between the value for N_2 (45.8 cal deg^{-1}) and that for F_2 (48.4 cal deg^{-1}). O_2 has an extra 2.2 cal mole^{-1} contribution to its entropy from the degeneracy of its electronic levels.

The two simple cases of entropy calculations which we have just performed, the entropy of a crystal at $0°K$ and the electronic contribution to the entropy in O_2, point the way to a general method for evaluating S in more complicated cases. The various contributions, translational, rotational, vibrational, and occasionally electronic, are evaluated separately. In evaluating each contribution, the exact values of all the energy levels of a given type must first be known. From these energies we can then calculate the population of each level using the Boltzmann Law. The population figures in turn enable us to evaluate W_{max}. The entropy contribution from this particular type of energy can then be calculated from equation (6.2). Finally, the various contributions can be summed to give a total value of the entropy.

The details of such calculations are unfortunately beyond the scope of this book. Nevertheless, some results can be quoted. It is possible, for instance, to calculate the *translational contribution* to the entropy of an ideal gas. The translational energy levels are first calculated by assuming that a molecule behaves with respect to translation like a particle in a three-dimensional rectangular box. The results are similar to those we obtained for the one-dimensional case in Chapter 3. The formula derived for the translational contribution to the entropy of one mole of gas from such a model is called the *Sackur-Tetrode* equation. It reads as follows:

$$S_{trans} = R \left[\log_e \frac{(2\pi mkT)^{3/2}}{h^3 N} V + \frac{5}{2} \right] \tag{6.12}$$

where m is the mass of a molecule of the gas, V is the molar volume, and h is Planck's constant.

Equation (6.12) readily simplifies to a more usable form if all the constants are collected together and evaluated. We then find that

$$S_{trans} = \left(\frac{3}{2}\right) R \log_e M + \left(\frac{3}{2}\right) R \log_e T - R \log_e P - 2.315 \text{ cal deg}^{-1}$$

$$\text{(6.13)}$$

$$= 6.865 \log_{10} M + 11.44 \log_{10} T - 4.576 \log_{10} P - 2.315 \text{ cal deg}^{-1}$$

where M is the molecular weight of the gas and P is its pressure in atmospheres.

Equation (6.13) enables us to calculate translational contributions to the entropy such as appear in Table 6-2, and also to calculate the standard molar entropies of monatomic gases such as the noble gases. As an example, let us calculate S_{298}^0 for argon.

We have

$$S_{trans} = (6.865 \log 39.95 + 11.44 \log 298.2 - 2.315) \text{ cal deg}^{-1}$$

$$= (11.00 + 28.32 - 2.315) \text{ cal deg}^{-1}$$

$$= 37.00 \text{ cal deg}^{-1}$$

Note that the qualitative behavior of equation (6.12) is exactly what we would expect. The heavier a molecule and the larger the volume occupied, the more closely spaced the translational levels should be. S should thus increase with both m and V, as indeed it does. Notice too, that the translational entropy increases with temperature.

The rotational and vibrational contributions to the entropy of simple gases can be deduced using procedures very similar to that used to calculate the translational entropy. There is one important difference, however. Rotational and vibrational energy levels cannot be calculated from simple properties such as mass and volume, but must be obtained experimentally from the *spectrum* of the gas. Examples of levels obtained in this way for diatomic gases have already been given in Figures 3-6 and 3-9. Once such energies are known, the general procedure outlined above can be followed to calculate the entropy. Some results of such calculations have already been given in Table 6-2.

In general, it is possible to calculate S from a knowledge of energy levels only for *simple gases*. In the case of solids and liquids the exact values of the energy levels are not known. Even in the case of gases, if the molecule is at all complex, it becomes difficult to find out what its vibrational energy levels are.

Calorimetric Values of the Entropy

As mentioned above, the second major method of estimating entropies depends upon the measurement of heat changes. When a system changes from one state to another, the corresponding change in entropy can be evaluated from a knowledge of how much heat energy the system absorbs and the temperature at which that change takes place.

It is not too difficult to establish the relationship between heat change and entropy for a change occurring at *constant volume*. Relationships for changes at constant pressure, and others in which both pressure and volume change, are more difficult to derive.

Suppose the system in which we are interested is constrained at constant volume, and there is no possibility of it doing electrical work, radiating heat energy, or otherwise losing energy. If the system is heated slightly, the first law of thermodynamics, equation (5.6), tells us that

$$\Delta E = q - w$$

but since the system is held at constant volume, no expansion work can be done; i.e., $w = 0$, so that:

$$\Delta E = q \qquad \text{(6.14)}$$

Equation (6.14) points to the fact that, since no expansion work is done, any heat absorbed by the system results only in the increase of the internal energy; that is, in the excitation of particles from lower to higher energy levels. Notice also that since the process occurs at constant volume, the constraint on the particles is not altered. As a result, the energy levels are not shifted and their energies remain constant.

Let us now follow the excitation of just one molecule from level i, to level j, where j is higher in energy than i. The heat energy absorbed, q_m, will be equal to the difference in energy between the two levels,

$$q_m = \Delta E = \epsilon_j - \epsilon_i \qquad \text{(6.15)}$$

(q is subscripted with an m to remind us that only one molecule is involved).

Let us now see how the entropy is changed by this excitation of one molecule. Initially the populations of the two levels will be n_i and n_j. For State 1, then, we have:

$$W_1 = \frac{n!}{n_0! \, n_1! \, n_2! \ldots n_i! \ldots n_j! \ldots}$$

After the jump, only two levels will have altered populations: level i will be depleted by one and will have a new population of $(n_i - 1)$, while level j will have its population increased by one to $(n_j + 1)$. For State 2, then, we have:

$$W_2 = \frac{n!}{n_0! \, n_1! \, n_2! \ldots (n_i - 1)! \ldots (n_j + 1)! \ldots}$$

We can now take the ratio of W_2 to W_1. All the terms cancel except those involving levels i and j. We find:

$$\frac{W_2}{W_1} = \frac{n_i! \, n_j!}{(n_i - 1)! \, (n_j + 1)!}$$

Remembering that $n! = n \times (n - 1)!$, this simplifies the relation to

$$\frac{W_2}{W_1} = \frac{n_i}{n_j + 1}$$

Since n_j is so large, we can make the approximation:

$$\frac{W_2}{W_1} = \frac{n_i}{n_j} \tag{6.16}$$

Using the Boltzmann law to express n_i and n_j in terms of n_0, equation (6.16) becomes:

$$\frac{W_2}{W_1} = \frac{n_0 \, e^{-\epsilon_i/kt}}{n_0 \, e^{-\epsilon_j/kt}}$$

$$= e^{(\epsilon_j - \epsilon_i)/kT}$$

Taking natural logarithms, we find:

$$\log_e\left(\frac{W_2}{W_1}\right) = \frac{\epsilon_j - \epsilon_i}{kT}$$

or

$$k \log_e W_2 - k \log_e W_1 = \frac{\epsilon_j - \epsilon_i}{T}$$

From equation (6.14) and the definition of entropy, we then have:

$$S_2 - S_1 = \frac{q_m}{T}$$

or

$$\Delta S_m = \frac{q_m}{T} \tag{6.17}$$

where ΔS_m is the entropy change occurring for the transfer of just *one* molecule from one level to another.

We can now extend our argument to a real process occurring in a system held at constant volume. Let the entropy change in this real process be ΔS and the heat energy absorbed be q. Such a process will involve the transfer of a very large number of molecules between levels, but we can argue each of these transfers, one at a time, in exactly the same way as above. Adding up all such entropy changes, we then obtain:

$$\Sigma \Delta S_m = \Sigma \frac{q_m}{T} \qquad (6.18)$$

Now the sum of all the single-transfer entropy changes is equal to the total entropy change of the real process, or

$$\Delta S = \Sigma \Delta S_m \qquad (6.19)$$

Furthermore, if (and only if) the process occurs at *constant temperature,* we. can write:

$$\Sigma \frac{q_m}{T} = \frac{1}{T} \Sigma q_m = \frac{q}{T} \qquad (6.20)$$

Inserting equations (6.17) and (6.18), we then obtain:

$$\Delta S = \frac{q}{T} \qquad (6.21)$$

which is our final result. Equation (6.21) tells us that for a change at constant volume and at *constant temperature,* the increase in entropy is equal to the total heat energy absorbed by the system divided by the temperature.

Although it is beyond our ability to prove here, equation (6.21) applies to *any* change at constant temperature whether the volume be kept constant (and the energy levels unshifted) or not. There is only one proviso. Throughout the change, the system must be in a Boltzmann distribution. If it is not, we cannot use the Boltzmann Law to derive equation (6.17).

As an example of the use of equation (6.21) to calculate an entropy change, let us calculate ΔS for the vaporization of one mole of water at constant temperature and pressure.

$$H_2O(l) \rightarrow H_2O(g) \quad (100°C, 1 \text{ atm})$$

The heat absorbed by this process is found experimentally to be 9.72 kcal. Thus q = 9,720 cal and T = 373°K. The use of equation (6.21) then yields the result:

$$\Delta S = \frac{9,720}{373} = 26.1 \text{ cal deg}^{-1}$$

Equation (6.21) enables us to calculate entropy changes only for processes occurring at constant temperature. It is possible to develop the same theory to encompass changes in which the temperature is not constant. Such a development, however, requires the use of the integral calculus.*

Suffice it to say that it is possible to evaluate the entropy of a substance by assuming it to have a zero entropy at $0°K$ and then calculating the entropy change which occurs as we heat it up from the solid state at $0°K$ to the temperature required, usually $25°C$. If the substance is a solid at $25°C$, we need only know its molar heat capacity (which is usually not constant) as a function of temperature over the range $0°K$ to $298°K$ ($25°C$). In the case of liquids and gases, we must also make allowances for changes in entropy which accompany melting and vaporization by means of calculations similar to that just done for water, using equation (6.21).

The calorimetric values for the entropy obtained in this way for simple gases usually agree, within the limits of experimental error, with the spectroscopic values. Examples of entropy values determined by both methods are given in Table 6-3. If one considers how different the two methods of obtaining these entropy values are, the agreement between them is a very impressive confirmation of the validity of both methods, and of the theory underlying them.

Table 6-3

Comparison of spectroscopic and calorimetric values of the standard molar entropy of various gases at $25°C$.

Gas	Standard Molar Entropy $(cal \ deg^{-1} \ mole^{-1})$	
	Spectroscopic	Calorimetric
Ne	34.95	35.01
Ar	36.99	36.95
Cl_2	53.31	53.31
HCl	44.65	44.47
N_2	45.77	45.93
H_2S	49.10	49.10
CO_2	51.05	51.11
NH_3	45.98	45.96
CH_4	44.47	44.46

---

*Readers who are familiar with the integral calculus will readily recognize that if S_m and q_m are both very small, equation (6.18) can be rewritten as

$$\Delta S = \int_1^2 dS = \int_1^2 \frac{dq}{T}$$

SUMMARY

In this chapter, a quantity called the entropy, whose symbol is S, has been introduced. The entropy is a quantity which increases with the number of occupied energy levels and hence, under standard conditions, with the spacing of the levels. In simple cases it is possible to tell, merely by inspection, which of two substances has the higher standard molar entropy and whether the entropy will increase or decrease in a reaction.

Entropy is both a state function and an extensive property. This makes it a quantity easy to manipulate. Values of S can be added and subtracted like values of E as considered in the previous chapter.

The entropy of a substance cannot be measured directly. Values of S can be found either by calculations based on energy level data, or by computation from data derived from experimental studies of heat changes.

PROBLEMS 6

1. From each of the following pairs of compounds, choose the one with the higher standard molar entropy.
   a. $Ar(g)$, $Kr(g)$
   b. $H_2(g)$, $HCl(g)$
   c. $HBr(g)$, $HI(g)$
   d. $CH_4(g)$, $CF_4(g)$
   e. $H_2S(g)$, $HCl(g)$
   f. $NaF(s)$, $MgO(s)$ (both have the same crystal lattice)
   g. $NaNO_3(s)$, $MgCO_3(s)$ (both have the same crystal lattice)

2. Although NH_3 is a lighter molecule than N_2, the standard molar entropy of NH_3 gas is larger than that of N_2 gas. Explain why.

3. Two of the following S^0 values are incorrectly given. Say which they are without consulting any tables.

Gas	S^0_{298} (cal deg^{-1})
N_2	45.8
CH_4	74.5
BF_3	60.7
SO_2	59.3
O_3	57.1
N_2O_4	72.7
I_2	27.8
H_2S	49.2

4. Without consulting tables, arrange the following sets of substances in order of increasing standard molar entropy.
   a. $F_2(g)$, $H_2(g)$, $Cl_2(g)$, $I_2(g)$, $Br_2(g)$
   b. $CH_3I(g)$, $CH_3D(g)$, $CH_3F(g)$, $CH_3Br(g)$, $CH_3Cl(g)$
   c. $S_2(g)$, $N_2(g)$, $H_2(g)$, $F_2(g)$
   d. CaS, KCl, TiN, CaO, KI (all of these solids crystallize with the NaCl type lattice.)
   e. $He(g)$, $D_2O(g)$, $Ne(g)$, $ND_3(g)$, $HF(g)$
   f. $S_2Cl_2(g)$, $H_2S(g)$, $H_2O(g)$, $H_2O_2(g)$

5. From Table 6-1, calculate ΔS^0 for the following reactions.
   a. $H_2(g) + Cl_2(g) \rightarrow 2HCl(g)$
   b. $2N_2(g) + 3H_2(g) \rightarrow 2NH_3(g)$
   c. $N_2(g) + O_2(g) \rightarrow 2NO(g)$
   d. C (diamond) $+ O_2(g) \rightarrow CO_2(g)$
   e. 2C (diamond) $+ O_2 \rightarrow 2CO(g)$

6. Without consulting tables, predict whether the entropy change will be positive, negative, or small in the following reactions:
   a. $H_2(g) + O_2(g) \rightarrow 2H_2O(l)$
   b. $2CO(g) + O_2(g) \rightarrow 2CO_2(g)$
   c. $H_2O(g) + C(s) \rightarrow CO(g) + H_2(g)$
   d. $CH_4(g) + 2O_2(g) \rightarrow CO_2(g) + 2H_2O(g)$
   e. $4HCl(g) + O_2(g) \rightarrow 2Cl_2(g) + 2H_2O(g)$
   f. $H_2O(l) + CuO(s) \rightarrow Cu(OH)_2(s)$
   g. $2KClO_3(s) \rightarrow 2KCl(s) + 3O_2(g)$
   h. $CH_4(g) + Cl_2(g) \rightarrow CH_3Cl(g) + HCl(g)$

7. A precise value of S_{298}^0 for $NCl_3(g)$ has never been determined. Make an intelligent guess at its value by contrasting S^0 values in Appendix 3 for various similar gases.

8. What is ΔS^0 for the reaction

$$A(g) \rightarrow B(g)$$

if A and B are optical isomers?

9. Show why a substance cannot have a negative value of S.

10. If $W_{max} = (10)^{10^{24.5}}$, calculate S.

11. If the value of S for a system is 50 cal deg^{-1}, what is the value of W_{max} for the system?

12. Liquid methane boils at $112°K$ and one atmosphere pressure. The heat energy absorbed is 2.22 kcal per mole of CH_4. Calculate ΔS for the process

$$CH_4(l) \rightarrow CH_4(g) \quad (112°K, 1 \text{ atm})$$

13. Liquid sodium boils at $899°C$ and one atmosphere pressure. The heat energy absorbed is 23.7 kcal per mole of sodium. Calculate ΔS for

the process

$$Na(l) \rightarrow Na(g) \quad (899°C, 1 \text{ atm})$$

14. Calculate the standard molar entropy of neon at $25°C$ using the Sackur-Tetrode equation.

15. Calculate the translational contribution to the entropy for O_2 at $25°C$. From this figure and the S^0 value for O_2 at $25°C$ (49.0 cal deg^{-1}), find the rotational contribution to the entropy. Assume that the vibrational contribution is negligible. Do not forget the electronic contribution to the entropy.

16. The molecule NO has an unpaired electron which causes it to have a doubly degenerate ground state. Calculate the electronic contribution to the entropy for this gas.

17. When CO is solidified, the molecules in the crystal appear to be randomly oriented, i.e.

CO CO OC CO

OC OC CO CO

OC CO OC OC

OC OC OC CO

rather than regularly oriented,

CO CO CO CO

CO CO CO CO

CO CO CO CO

CO CO CO CO

and this random orientation persists down to $0°K$. What would be the entropy of a mole of solid CO at $0°K$ under such circumstances?

18. Suppose that instead of defining the entropy by use of the formula

$$S = k \log_e W_{max}$$

we had defined it by

$$Z = \log_{10} W_{max}$$

What would the relationship between ΔZ and q have been for a constant-temperature, constant-volume process equivalent to equation (6.21)?

19.　A system consists of a mole of particles. Two thirds of these are in the ground state, and one third are in the first excited state. No other levels are occupied. Use the Stirling approximation to calculate the entropy of this system.

20.　Figure 4-7 (iii) (p. 85) shows the energy levels of two hypothetical isomers A and B. Both have the same ground state energies, but the levels of isomer A are spaced at an interval which is twice that for isomer B. Show that for the reaction

$$A(g) \to B(g)$$

ΔS is zero at $0°K$ and approaches the limit $R \log_e 2$ at high temperatures. Deduce the behavior of ΔS at very low and very high temperatures for the four other cases shown in Figure 4-7.

21.　Explain why the S^0 value for gaseous CH_4 is *less* than that for gaseous NH_3.

Chapter 7

ENERGY
AND
ENTROPY
IN
ACTION

INTRODUCTION

The previous two chapters have served to introduce us to two thermo-dynamic functions, the internal energy, E, and entropy, S. The introduction of these two functions puts us in a position to return to the subject of chemical equilibrium and to develop it beyond the energy-level picture discussed in Chapter 4. Now that we know something about the internal energy and the entropy, we can begin to treat chemical equilibrium entirely in terms of these two functions and abandon energy levels altogether. Such an approach will enable us to make predictions about the behavior of a very large number of equilibria.

A SIMPLE EXAMPLE

In order to see how E and S can be applied to chemical equilibrium, let us first consider a simple example. This is the equilibrium between the two gaseous isomers with the formula C_5H_{10}, which we discussed in the last chapter and which are illustrated in Figure 6-7:

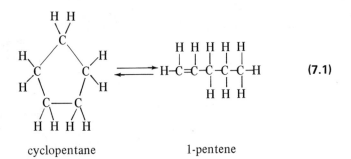

cyclopentane   1-pentene

We can estimate an approximate ΔE^0 for reaction (7.1) from bond energy tables*. The number of C–H bonds is the same for both isomers, so the making and breaking of these bonds makes no contribution to ΔE^0. The situation in the case of carbon-carbon bonds is different, though. Cyclopentane contains five C–C bonds, while 1-pentene contains three C–C bonds and one C=C bond. The formation of 1-pentene from cyclopentane thus corresponds to the replacing of *two* carbon-carbon single bonds by *one* carbon-carbon double bond. From Table 5-3, we find that

$$\Delta E^0 = 2 \, D_{C-C} - D_{C=C}$$

$$= 2(83) - 147 = +19 \text{ kcal}$$

The exact value derived from direct measurements is +13.5 kcal.

What does the fact that this reaction is endothermic to the extent of 13.5 kcal per mole mean in terms of energy levels? It means that the ground state of the cyclopentane molecule is much lower than the ground state of the 1-pentene molecule, the difference being about 13.5 kcal per mole. When one mole of cyclopentane molecules is converted into a mole of 1-pentene molecules, the major change in energy which occurs is a change in *bonding,* that is, a change in *electronic energy.* Changes in translational, rotational, and vibrational energies are very small by comparison. If the ground state energies are separated by some 13.5 kcal of energy, this means that one set of energy levels is much lower-lying than the other. The value of ΔE^0 thus tells us a great deal. Once we know that ΔE^0 is +13.5 kcal, *we also know that the set of energy levels corresponding to cyclopentane is lower-lying than the set of levels corresponding to 1-pentene.* Given this fact, we can use the energy level approach of Chapter 4 to predict that in equilibrium (7.1), *cyclopentane will be favored over 1-pentene at low temperatures.* In other words, the value of ΔE^0 tells us how the equilibrium under discussion will behave at low temperatures.

---

*The superscript zero in ΔE^0 has the same significance as in ΔS^0. ΔE^0 refers to the change in internal energy when reactants and products are all at the same standard pressure of one atmosphere.

By contrast, a consideration of the *entropy* changes accompanying reaction (7.1) tells us how it will behave at *high temperatures*. The entropy of the two isomers involved in the reaction has already been considered in Chapter 6. There, you will remember, we saw that 1-pentene has a larger entropy than cyclopentane, mainly because its easily bent chain of carbon atoms makes for a larger vibrational contribution to the entropy than is found for the rigid ring structure of cyclopentane. Such a conclusion can be checked from Table 6-1. S^0 for 1-pentene is 83.1 cal deg^{-1}, while the value for cyclopentane is 70.0 cal deg^{-1}. ΔS^0 for reaction (7.1) is thus +13.1 cal deg^{-1}.

What significance do these values of S^0 have for a discussion of the equilibrium behavior of reaction (7.1)? They not only tell us which isomer has the higher standard molar entropy, but they also tell us which side of the equilibrium corresponds to the *more closely-spaced levels*. This in turn indicates which side will be favored at high temperatures. In other words, once we know that ΔS^0 for the reaction is +13.1 cal deg^{-1}, we also know that the standard molar entropy of 1-pentene is higher than that of cyclopentane, and that in consequence it is 1-pentene that will be the favored isomer at high temperatures.

THE LOW ENERGY – HIGH ENTROPY RULE

The arguments we have just applied to equilibrium (7.1) apply in general to all isomerization equilibria. For any equilibrium of the type

$$A(g) \rightleftharpoons B(g)$$

we can always identify that isomer which has the *lower-lying levels* as being the isomer with the lower value of internal energy. We can likewise identify that isomer which has the *more closely spaced levels* as the isomer which has the *higher value of the standard molar entropy*. Having established this principle, it is clear that all we need to be able to predict the direction of an isomerization reaction is *thermodynamic* information. We need the value of ΔE^0 to tell us which isomer has the lower energy and the value of ΔS^0 to tell us which isomer has the higher entropy.

Similar remarks apply to gaseous reactions in general. Whether we are considering an isomerization reaction or a more complicated reaction, ΔE^0 always tells us in which direction the reaction will go at low temperatures, while ΔS^0 tells us in which direction it will go at high temperatures. This dependence of the direction of a chemical reaction on the energy and entropy is best described in terms of a rule. In this book such a rule will be referred to as the *Low Energy-High Entropy Rule*. At the moment we are not in a position to prove this rule. We shall take it as axiomatic.

The low energy-high entropy rule reads as follows:

THAT SIDE OF A CHEMICAL EQUILIBRIUM WITH THE LOWER INTERNAL ENERGY IS FAVORED AT LOW TEMPERATURES, WHILE THAT SIDE OF A CHEMICAL EQUILIBRIUM WITH THE HIGHER ENTROPY IS FAVORED AT HIGH TEMPERATURES.

In using this rule, the terms "low energy side" and "high entropy side" must be interpreted in terms of ΔE^0 and ΔS^0. If ΔE^0 for the reaction being considered is positive, we take it that the reactants correspond to the lower energy side of the reaction, while if ΔE^0 is negative then it is the products which have the lower energy. Similar remarks apply to ΔS^0. If this quantity is positive, then the products must be regarded as constituting the high entropy side of the reaction, while if ΔS^0 is negative, the reactants have the higher entropy.

As an example of how the low energy-high entropy rule may be applied, consider the reaction:

$$4HCl(g) + O_2(g) \rightarrow 2H_2O(g) + 2Cl_2(g) \qquad (7.2)$$

It is known from experimental data that $\Delta E^0 = -26.6$ kcal for this reaction, while from Table 6-1, $\Delta S^0 = -30.6$ cal deg^{-1}.

Since ΔE^0 is negative, the products of reaction (7.2) are lower in internal energy than the reactants. Application of the low energy-high entropy rule then leads to the prediction that the *products*, H_2O and Cl_2, will be favored at *low* temperatures. Similarly, the fact that ΔS^0 for this reaction is negative means that the reactants, HCl and O_2, have the higher entropy. The rule would thus predict that at *high* temperatures the *reactants* will be favored over the products.

Experimentally, reaction (7.2) behaves in the manner we have predicted. The reaction can be made to go in either direction. The forward reaction corresponds to an outdated method for preparing chlorine gas, called the Deacon process. The reverse reaction, which occurs only at higher temperatures, has been used for preparing HCl of high purity.

The low energy-high entropy rule assumes that it is valid to talk about one side of a reaction as being the "low energy side" without specifying the temperature. In other words, it assumes that if ΔE^0 is positive at one temperature it will also be positive at any other temperature. Such an assumption is valid because, as we have noted several times, ΔE^0 is determined mainly by changes in bonding, i.e., by changes in electronic energy. Although changes in other forms of energy also occur, these are almost always small, seldom amounting to more than a few kilocalories per mole. This being the case, we expect ΔE^0 to vary by no more than a few kilocalories over the range of temperatures normally encountered in chemical reactions.

In Table 7-1 the variation of ΔE^0 with temperature is shown for five different gaseous reactions in the range 0^0K to 1000^0K. In no case is the variation in ΔE^0 larger than 5 kcal over this temperature range. Consider, for instance, the second reaction featured in the table; the decomposition of N_2O_4 to NO_2. At $0°K$, ΔE^0 for this reaction is $+12.7$ kcal. The value corresponds to the energy required to decompose a mole of N_2O_4 molecules, all in their ground states, into two moles of NO_2 molecules, all in their ground states. In other words, 12.7 kcal corresponds to the ground state-to-ground state change in *electronic energy*. As the temperature rises above $0°K$, the value of ΔE^0 changes from this initial value, first increasing slightly and then decreasing. These changes, which reflect differences in vibrational, rotational, and translational energies, are never more than 3 kcal mole^{-1}.

Table 7-1

Variation of ΔE^0 and ΔS^0 with temperature
for several gaseous reactions.

$T(K°)$	ΔE^0 (kcal)	ΔS^0 (cal deg^{-1})
	$H_2(g) + I_2(g) \rightarrow 2HI(g)$	
0	-2.0	0.0
100	-2.1	$+6.1$
298	-2.3	$+5.2$
500	-2.7	$+4.3$
1000	-3.2	$+3.5$
	$N_2O_4(g) \rightarrow 2NO_2(g)$	
0	$+12.7$	0.0
100	$+13.2$	$+39.6$
298	$+13.1$	$+42.0$
500	$+12.3$	$+41.0$
1000	$+9.7$	$+38.9$
	$N_2(g) + 3H_2(g) \rightarrow 2NH_3(g)$	
0	-18.7	0.0
100	-19.7	-36.9
298	-20.9	-47.5
500	-22.1	-52.6
1000	-23.0	-56.9
	$2SO_2(g) + O_2(g) \rightarrow 2SO_3(g)$	
0	-45.8	0.0
100	-46.3	-39.9
298	-46.8	-44.9
500	-46.4	-45.2
1000	-44.7	-44.3
	$CO(g) + NO_2(g) \rightarrow CO_2(g) + NO(g)$	
0	-53.9	0.0
100	-53.9	-2.3
298	-54.0	-3.1
500	-53.9	-2.9
1000	-53.6	-2.5

An example worth noting in Table 7-1 is the first reaction, the formation of HI from its elements. For this reaction the ground state-to-ground state energy difference is only 2.0 kcal per mole of H_2. Very few gaseous reactions involve energy differences as small as this. Yet even though ΔE^0 is so small, it still does not change sign with temperature.

The entropy changes listed in Table 7-1 are also worth commenting on. The behavior of ΔS^0 is similar, though not identical, to that of ΔE^0. The chief difference is that ΔS^0 is zero for all reactions at $0°K$. This is in agreement with the Third Law of Thermodynamics (Chapter 6). If all substances have zero energy at absolute zero, there can be no change in entropy when some substances change into others.

Notice that in none of the reactions listed in the table does ΔS^0 change sign with temperature. There is a simple explanation for this. A closely spaced set of levels will always have more levels occupied than a widely-spaced set at the same temperature. The actual temperature is unimportant. A movement from a less closely spaced to a more closely spaced set of levels will thus always correspond to an increase in entropy, no matter what the temperature.

The approximation is sometimes made that the value of ΔS^0 for a reaction does not vary with temperature. In the vicinity of $0°K$, this is obviously a bad approximation. Between $100°K$ and $1000°K$, however, the approximation improves, as Table 7-1 shows. Nevertheless it is more nearly valid to say that ΔE^0 is constant than it is to say that ΔS^0 is constant over the range of temperatures given in the table.

TWO COMPOUNDS OF NITROGEN

In order to familiarize ourselves with the low energy-high entropy rule, we will spend most of the rest of this chapter considering examples of its use. Hopefully, these examples will also increase our insight into, and understanding of, chemical equilibrium.

As our first example, consider the equilibrium

$$N_2(g) + 3H_2(g) \rightleftharpoons 2NH_3(g) \quad (1 \text{ atm}) \qquad (7.3)$$

We must first determine which side of reaction (7.3) is the low energy side. The reaction corresponds to the breaking of one $N\equiv N$ bond and three $H\!-\!H$ bonds and the making of six N–H bonds. From Table 5-3 we then estimate ΔE^0:

$$\Delta E^0 = D_{N\equiv N} + 3\,D_{H-H} - 6\,D_{N-H}$$

$$= 226 + 3(104) - 6(93) = -20 \text{ kcal}$$

A comparison with the accurate values given in Table 7-1 shows that this is within 3 kcal of the correct value over the range from $100°K$ to $1000°K$.

The basic reason reaction (7.3) is exothermic is that it corresponds to the making of *six* bonds at the expense of breaking *four*. Even though the bonds which are broken are individually stronger than those which are made, this effect is not enough to offset the greater number of bonds formed. As a consequence, it is the *right-hand* side of reaction (7.3) which is the low energy side. Accordingly, we predict that the equilibrium position will favor ammonia rather than its elements at low temperatures.

Next, let us determine which side of reaction (7.3) has the higher entropy. We use the rule that the side of a reaction which has the larger number of molecules also has the higher entropy. This is, of course, the *left-hand side* of (7.3) (4 molecules versus 2). This result is confirmed both by Table 6-1 and by Table 7-1. From the latter table we see that ΔS^0 for this reaction is of the order -50 cal deg^{-1} over a wide range of temperatures. Such a result leads us to predict that at high temperatures ammonia will decompose into its elements.

The prediction that ammonia is formed at low temperatures but decomposes into its elements at high temperatures is amply vindicated by experiment. From experimental results it is found that at 1 atmosphere and 600°C, NH_3 is 99.8 per cent dissociated into its elements at equilibrium. As the temperature is lowered but the pressure maintained at 1 atmosphere, the percentage of dissociation decreases. It is 97.5 per cent at 400°C, 76.5 per cent at 200°C, and 16.4 per cent at 25°C.

The equilibrium between ammonia and its elements is of considerable industrial importance. Ammonia has been manufactured from its elements on a very large scale for more than fifty years. This is why the equilibrium has been so carefully studied experimentally — more so, perhaps, than any other gaseous equilibrium. It is noteworthy that in this reaction the equilibrium state cannot be achieved without a catalyst. However, as we have argued in Chapter 1, the presence of a catalyst does not affect the position of equilibrium. It merely facilitates its attainment.

The next example of an application of the low energy-high entropy rule is one which is very similar *on paper* to the ammonia equilibrium just considered. In practice, though, it behaves very differently. The reaction to be considered is that between nitrogen and chlorine:

$$N_2(g) + 3Cl_2(g) \rightleftharpoons 2NCl_3(g) \quad (1 \text{ atm}) \qquad (7.4)$$

ΔE^0 for this reaction can be estimated from Table 5-3 in a manner similar to that used in the previous example. We find that

$$\Delta E^0 = D_{N \equiv N} + 3 D_{Cl-Cl} - 6 D_{N-Cl}$$

$$= 226 + 3(58) - 6(48) = 112 \text{ kcal}$$

In contrast to the reaction between nitrogen and hydrogen, this reaction is *endothermic*. Again six bonds are made at the expense of breaking four. The major difference between the two reactions is that the six bonds which are formed here are N—Cl bonds rather than N—H bonds. Nitrogen and chlorine are fairly similar in electronegativity, so we do not expect them to form a very strong bond. At 48 kcal mole^{-1}, the N—Cl bond is a rather weak single bond in comparison to the 93 kcal mole^{-1} of the N—H bond. The formation of six such weak bonds is not enough to outweigh the energy absorbed when one very strong N≡N bond of 226 kcal mole^{-1} and three Cl—Cl bonds of 58 kcal mole^{-1} each are broken.

So much for the energy change in reaction (7.4). How about the entropy change? For this reaction ΔS^0 must certainly be negative by the same argument used for the ammonia reaction. Since the number of molecules decreases, so must the entropy. We cannot cross-check on this result from tables since the standard molar entropy of NCl_3 is not known. The reason for such a state of affairs will soon be apparent.

Now that the sign of both ΔE^0 and ΔS^0 have been estimated, we can apply the low energy-high entropy rule to reaction (7.4). Since the reactants are both the low energy and the high entropy side of the reaction, we expect them to be favored both at high and at low temperatures; that is, we expect nitrogen and chlorine to show almost no tendency to combine to form NCl_3 at any temperature. Such a prediction is borne out experimentally. No one has yet succeeded in making NCl_3 directly from its elements. Although the compound can be prepared in other ways, it is a notoriously unstable and dangerous substance. It shows a marked tendency to decompose explosively into its elements without provocation. It is for this reason that no one has yet dared to perform the thermal or spectroscopic measurements necessary to determine the value of its standard molar entropy.

It is as well at this point to emphasize the fact that thermodynamics tells us nothing about the *rate* at which an equilibrium can be established. The best it can do is to tell us where the equilibrium position is. In the two examples we have just discussed, we were able to predict correctly whether reactants or products were favored at a high or a low temperature. We were *not* able to predict that in one case the rate at which equilibrium is attained is so slow that a catalyst is required, whereas in the other the rate is so fast that an explosion results.

DOUBLE BONDS VERSUS SINGLE BONDS

The next example of the use of the low energy-high entropy rule which we will consider is an important type of reaction called *polymerization.* In a polymerization reaction, a large number of simple molecules called *monomers* are induced under the influence of heat or a catalyst to join up with each other to form a giant molecule or *polymer.* Perhaps the simplest polymer to consider is polyethylene, which is the familiar white waxy plastic from which "squeeze bottles" are made. Polyethylene has the molecular structure:

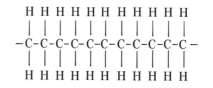

The carbon-carbon chain is some 1500 atoms long. Polyethylene can be polymerized from the gaseous monomer, $CH_2=CH_2$, with the aid of a catalyst. We shall write such a reaction as:

$$n \quad \begin{array}{c} H \quad H \\ | \quad | \\ C{=}C \\ | \quad | \\ H \quad H \end{array} \rightarrow \left\{ \begin{array}{c} H \quad H \\ | \quad | \\ -C-C- \\ | \quad | \\ H \quad H \end{array} \right\}_n \qquad \textbf{(7.5)}$$

where n is about 750.

The entropy change for reaction (7.5) is obviously negative, because of the large reduction in the number of molecules. This means that at a sufficiently high temperature the monomer, ethylene, will be favored. The reaction is possible at low temperatures only because the polymer is the low energy side of the reaction. Let us see why this should be so in terms of bond energies.

Since the polymerization leaves all the C–H bonds unchanged, we can ignore them. We need only consider the changes in the carbon-carbon bonding. For each ethylene unit which polymerizes, a C=C bond is replaced by two C–C bonds. Using Table 5-3 we can now estimate the magnitude of ΔE^0 for reaction (7.5):

$$\Delta E^0 = D_{C=C} - 2\, D_{C-C}$$

$$= 147 - 2(83) = -19 \text{ kcal}$$

per mole of ethylene involved. The polymer thus corresponds to the *low energy side* of reaction (7.5). That it can form at low temperatures is in conformity with the low energy-high entropy rule.

Notice that the basic reason why ΔE^0 is negative and not positive for this particular reaction lies in the relative values of the bond energies of the C=C and C–C bonds. ΔE^0 is negative because the *bond energy of the carbon-carbon double bond* (147 kcal mole^{-1}) *is less than twice that of the carbon-carbon single bond* (83 kcal mole^{-1}). If this were not the case, ΔE^0 would be positive. It would then not be possible to polymerize ethylene at all, no matter what the temperature. Furthermore, even if we could make polyethylene by some other method, it would be unstable with respect to decomposition into its monomer at all temperatures.

The same argument accounts for the stability of other polymers containing long chains of carbon atoms. Examples are:

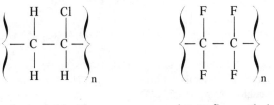

polyvinyl chloride   polytetrafluoroethylene

(P.V.C.)   (Teflon)

These polymers do not decompose into their monomers except at high temperatures because the process is endothermic. Basically, therefore, they also owe their stability to the relative values of the C=C and C—C bond energies.

Interestingly enough, the relative values of double and single bond energies in the case of nitrogen are the reverse of those of carbon. The bond energy of the N=N bond (100 kcal mole^{-1}) is *more* than twice the bond energy of the N—N bond (38 kcal mole^{-1}). This insures that any polymers containing long chains of nitrogen atoms will be unstable with respect to decomposition to the monomer. Consider, for instance, the polymerization reaction of the compound diimine, HN=NH:

$$n \quad \begin{array}{c} N{=}N \\ | \quad | \\ H \quad H \end{array} \rightarrow \left\{ \begin{array}{c} -N{-}N{-} \\ | \quad | \\ H \quad H \end{array} \right\}_n \tag{7.6}$$

Using the bond energy values just given, we find that

$$\Delta E^0 = 100 - 2(38) = 24 \, \text{kcal}$$

for each diimine unit. Such a result means that the left-hand side of equilibrium (7.6) is the low energy side. It is, of course, also the high entropy side. The monomer thus corresponds to the favored species at *all* temperatures. In agreement with this, we find that no one has yet succeeded in preparing a polymer of diimine.

If truth must be told, no one has yet succeeded in preparing diimine itself, either! Again, this is something that the combination of the low energy-high entropy rule and a further look at the energies of nitrogen-nitrogen bonds would lead us to expect. Diimine is unstable with respect to decomposition into nitrogen and hydrogen. A major reason for this state of affairs is the very high value for the bond energy of the triple bond in N_2. Consider the reaction

$$HN{=}NH(g) \rightarrow N{\equiv}N(g) + H_2(g) \tag{7.7}$$

ΔE^0 for this reaction can be estimated in the usual way from the bond energy data of Table 5-3. We find that

$$\Delta E^0 = D_{N=N} + 2 D_{N-H} - D_{N\equiv N} - D_{H-H}$$

$$= 100 + 2(93) - 226 - 104 \text{ kcal} = -44 \text{ kcal}$$

Since ΔE^0 is negative, the right hand side of equilibrium (7.7) is the low energy side. Because there are more molecules on that side of the equilibrium, it is also the high entropy side. Thus both the entropy and the energy factors favor the elements over the compound. This means that diimine is unstable with respect to decomposition into its elements at all temperatures.

We should notice that one of the important factors determining the sign of ΔE^0 in reaction (7.7), and hence the instability of diimine, is that the $N\equiv N$ bond is such a strong bond compared to the $N=N$ bond. Even though three bonds are broken and only two are made in reaction (7.7), the energy of the one $N\equiv N$ bond which is made is a thumping big 226 kcal mole^{-1}, quite enough to make the reaction exothermic.

The large value of the $N\equiv N$ bond has a similar effect in all reactions in which N_2 takes part or is evolved. In fact, it is very difficult to discuss the chemistry of nitrogen without introducing the value of this bond energy.

XENON DIFLUORIDE

When compounds of xenon were first prepared in 1962, many chemists were taken by surprise. Previous to this time, most chemists had taken it for granted that noble gases could not form compounds. Surprising or not, the preparation of such compounds in fact contradicts no thermodynamic principles. In what follows, we shall show that the preparation of XeF_2 is quite in accord with the low energy-high entropy rule.

The equation for the formation of xenon difluoride is as follows:

$$Xe(g) + F_2(g) \rightarrow XeF_2(g) \tag{7.8}$$

This reaction (like all reactions) is possible only if the entropy change is positive or the energy change is negative or if both conditions are met. Since the number of molecules decreases in the reaction, ΔS^0 is clearly negative. There is thus only one possibility left if reaction (7.8) is to occur at all. ΔE^0 must be negative.

This reaction corresponds to the breaking of one bond (the F–F bond) and the making of two bonds (both Xe–F bonds). Thus more bonds are broken than are made, a condition favorable to an exothermic reaction and a negative value for ΔE^0.

We can relate ΔE^0 to the values of bond energies by means of the equation

$$\Delta E^0 = D_{F-F} - 2\,(D_{Xe-F}) \tag{7.9}$$

Of the two bond energies featured in (7.9), only one can be found in Table 5-3. This is the F—F bond energy, whose value is given as 37 kcal mole^{-1}. The other bond energy has been determined experimentally from calorimetric measurements on both XeF_2 and other fluorides of xenon. The value obtained for the Xe—F bond energy in this way is 31 kcal mole^{-1}. Substituting these two bond energy values into equation (7.9), we find that

$$\Delta E^0 = 37 - 2(31) = -25 \text{ kcal}$$

The reaction is thus exothermic, and no thermodynamic principle is violated when it occurs.

There are two aspects of reaction (7.8) which are worth remarking on. The first is the effect on the reaction direction of the monatomic nature of Xe. If Xe were diatomic it would then be necessary to break a Xe—Xe bond in order to form XeF_2. The breaking of such a bond would make ΔE^0 less negative and the reaction less likely to occur. The fact that Xe is monatomic assists the formation of XeF_2 by reducing the number of bonds which must be broken.

The second point to notice is that although the Xe—F bond is a weak bond, this fact is not enough by itself to make XeF_2 unstable with respect to its elements. In the formation of XeF_2 only one weak bond, the F—F bond, must be broken. The two bonds which replace this one bond do not have to be very strong to make the reaction exothermic. From equation (7.9) we can calculate that even if the bond energy of the Xe—F bond were as low as 19 kcal mole^{-1} this would still be large enough to make ΔE^0 negative.

ADDITIONAL REMARKS

The examples of gaseous equilibria we have just considered are enough to demonstrate how useful the low energy-high entropy rule can be. It provides us with a quick qualitative prediction of the behavior of a large number of gaseous reactions. All we need know in order to apply the rule are the signs of the two quantities ΔE^0 and ΔS^0. Quite often we can predict these signs and apply the rule with no more information than a table of bond energies.

Despite the ease with which we can apply it, the low energy-high entropy rule has a very severe limitation. It is a qualitative rather than a quantitative rule. By using it we can, for instance, predict that in the equilibrium

$$N_2O_4(g) \rightarrow 2NO_2(g)$$

it is the dissociated species, NO_2, which is favored at high temperatures, and the associated species, N_2O_4, which is favored at low temperatures. But the rule tells us nothing about what percentage of N_2O_4 is dissociated at what temperatures. As a result we have no idea of whether the equilibrium state for the reaction is well to the right or well to the left at a given temperature.

In order to overcome this difficulty, we have no option but to make our treatment of thermodynamics quantitative rather than qualitative. Only then can we begin to answer questions about how far a reaction will go at a specified temperature. Most of the remainder of the book will be devoted to making thermodynamics quantitative in this way.

Before concluding this chapter, there is still one aspect of the low energy-high entropy rule which must be mentioned. Although up till now we have used it only for cases of gaseous equilibrium, the rule in fact applies quite generally to all equilibria, whether gases are involved or not. For instance, we can apply it to the equilibrium

$$H_2O(s) \rightleftarrows H_2O(l) \quad (1 \text{ atm}) \qquad \textbf{(7.10)}$$

As is well known, heat (the so-called latent heat of fusion) must be absorbed to convert ice into water. Ice must therefore be lower in internal energy, per mole, than water. On the other hand, because of the lower constraint on molecular motion in the liquid as opposed to the solid, liquid water must have a higher standard molar entropy than ice. Applying the low energy-high entropy rule, we predict, correctly, that liquid water is favored at high temperatures and solid ice at low temperatures.

Despite the correctness of our prediction, there is a very obvious difference between equilibrium (7.10) and a gaseous equilibrium such as

$$A(g) \rightleftharpoons B(g) \qquad \textbf{(7.11)}$$

where A and B are two isomers. Suppose the compound A in equilibrium (7.11) is the isomer of lower energy, while B is the isomer of higher entropy. As we raise the temperature from absolute zero the equilibrium will shift from 100 per cent A to something very close to 100 per cent B at very high temperatures. No such gradual transition from one side of the equilibrium to the other occurs in the case of the ice-water equilibrium as the temperature rises. Below $0°C$, only pure ice is stable, while above this temperature only pure water is stable. At exactly $0°C$ *any* ice-water mixture ranging from 100 per cent ice to 100 per cent water is a genuine equilibrium mixture.

This behavior of the ice-water equilibrium is not something which fits very easily into the picture of equilibrium developed in Chapter 4. If we are to regard such an equilibrium as a competition between two sets of energy levels for population, then the rules of the competition must be very different from those governing a gaseous equilibrium. They must be more of the "winner takes all" variety. Above the melting point the liquid is the winner, and below the melting point the solid is. Only at the melting point itself are both species able to coexist.

The truth is that the behavior of a solid-liquid equilibrium just described is difficult to interpret in terms of a molecular picture. On the other hand, once we abandon a molecular picture, and think purely in terms of thermodynamic functions, it becomes quite easy to handle something like the ice-water equilibrium. The "winner takes all" aspect of this equilibrium is simple to explain in terms such as ΔE^0 and ΔS^0, as we shall see in Chapter 9.

PROBLEMS 7

1. Using only a table of bond energies and the low energy-high entropy rule, predict the behavior of the following equilibria (i) at low and (ii) at high temperatures.
   a. $N_2(g) + 3F_2(g) \rightleftharpoons 2\,NF_3(g)$
   b. $N_2(g) + 3Br_2(g) \rightleftharpoons 2NBr_3(g)$
   c. $2H_2O_2(g) \rightleftharpoons 2H_2O(g) + O_2(g)$
   d. $H_2O_2(g) \rightleftharpoons H_2(g) + O_2(g)$
   e. $CO(g) + 2H_2(g) \rightleftharpoons CH_3OH(g)$
   f. $H_2(g) + H_2C{=}CH_2(g) \rightleftharpoons H_3C{-}CH_3(g)$
   g. $CH_4(g) + 2O_2(g) \rightleftharpoons CO_2(g) + 2H_2O(g)$
   h. $CH_4(g) + 2H_2O(g) \rightleftharpoons CO_2(g) + 4H_2(g)$
   i. $2CO_2(g) \rightleftharpoons 2CO(g) + O_2(g)$
   j. $4NH_3(g) + 3O_2(g) \rightleftharpoons 2N_2(g) + 6H_2O(g)$
   k. $3F_2O(g) + 4NH_3(g) \rightleftharpoons 2N_2(g) + 3H_2O(g) + 6HF(g)$

2. For the reaction

$$A(g) + B(g) \rightleftharpoons C(g) + D(g)$$

K_c has the value 213 at $25°C$ and 0.0036 at $800°C$. Decide whether reactants or products have the higher internal energy. Which is higher in entropy?

3. A chemical reaction has the form

$$A(g) + B(g) \rightleftharpoons C(g) + D(g).$$

Predict whether reactants or products will be favored at high or low temperatures if:
   a. Both ΔE^0 and ΔS^0 are positive.
   b. ΔE^0 is positive but ΔS^0 negative.
   c. ΔE^0 is negative but ΔS^0 positive.
   d. Both ΔE^0 and ΔS^0 are negative.

4. Calcite and aragonite are two different crystalline forms of $CaCO_3$. Calcite has a slightly higher molar entropy than aragonite, and also a slightly higher internal energy. Which form of $CaCO_3$ will be favored (i) at high temperatures and (ii) at low temperatures?

5. For the reaction

$$C(graphite) \rightarrow C(diamond) \quad (25°C, 1 \text{ atm})$$

ΔE^0 has the value +0.5 kcal while ΔS^0 has the value −0.8 cal deg^{-1}. Use this information to predict which form of carbon will be stable at (i) high temperatures and (ii) low temperatures.

6. Gray tin is stable below 18°C, while white tin is stable above this temperature. Which allotrope has (i) the lower standard molar internal energy and (ii) the lower standard molar entropy?

7. The vapor pressure of any liquid rises with temperature. Interpret this behavior in terms of the low energy-high entropy rule.

8. It is impossible to make a cement mixture which will set without releasing heat. Discuss this statement in terms of the low energy-high entropy rule.

9. Fluorine forms stable covalent compounds with almost all the non-metallic elements. Suggest why.

10. Suggest why nobody has yet prepared the compound H—O—O—O—O—O—H.

11. The compound difluorimine, HNF_2, is an unstable explosive material. Suggest why.

12. The compound XeO_3 is a dangerously unstable compound. Suggest why this should be so, given that the mean bond energy of the Xe—O bond is 21 kcal mole^{-1}.

Chapter 8

THE
ENTHALPY

INTRODUCTION

In our discussion up to this point, a very important aspect of the internal energy seems to have been neglected. Although exact measured values of the entropy have been presented in the form of tables, there has been no attempt to do the same for the internal energy. We have only been able to calculate *approximate* values of ΔE^0 from a table of bond energies.

The truth is that accurate internal energy values are seldom tabulated. What we find instead are tables involving a function called the *enthalpy*. The enthalpy, symbol H, is a quantity very closely related to the internal energy E. In this chapter we will learn about this third thermodynamic function and how to distinguish it from the internal energy.

THE MEASUREMENT OF HEAT CHANGES

In Chapter 5 we dealt with the subject of heat energy without considering how to measure the heat changes which accompany a reaction. The time has now come to make good this deficiency. Two representative examples of heat measurements will be considered. One example involves an exothermic process and the other an endothermic process.

The first example to be considered is the reaction

$$Mg(s) + \tfrac{1}{2}O_2(g) \rightarrow MgO(s) \tag{8.1}$$

The heat change which accompanies this reaction can be measured in an apparatus, called a *bomb calorimeter,* which is illustrated diagrammatically in Figure 8-1. The innermost part of this apparatus is a small dish into which a weighed sample of the magnesium (or other combustible substance) is placed. A fine iron wire, which heats up rapidly when an electric current is passed through it, is also placed in the dish, in contact with the magnesium sample. The dish and

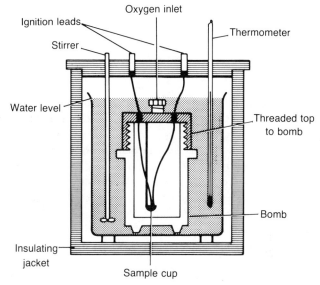

Figure 8-1. A bomb calorimeter.

its contents are enclosed in a heavy-walled steel bomb, and the cover is screwed on to give an air-tight fit. Oxygen gas from a cylinder is now used to fill the bomb at a pressure of some fifty atmospheres. The bomb is immersed in a weighed quantity of water, the water being surrounded by a thermally insulated container. The initial temperature of the water is measured with a sensitive thermometer.

The reaction is now initiated by passing an electric current momentarily through the fine iron wire. The sample ignites, attains a very high initial temperature, and then cools down by releasing heat to the bomb and the surrounding water. The rise in temperature of the water is measured by the sensitive thermometer.

The quantity of heat evolved could be calculated from the mass of the water, the mass and the specific heat of the steel bomb, and the rise in temperature. More commonly, the "heat capacity" of the calorimeter is used instead. The heat capacity of the calorimeter is the amount of heat energy absorbed by the bomb and water when the temperature is increased by 1°C.

In a typical experiment, 0.500 g of magnesium was ignited in an excess of oxygen and the temperature rise was found to be 2.03°C. The heat capacity of the calorimeter was known to be 1.46 kcal deg^{-1}.

The heat capacity for the calorimeter tells us that the release of 1.46 kcal of heat energy causes a rise of 1°C. The amount of heat evolved in the experiment (which causes a rise of 2.03°C) must thus be

$$2.03 \text{ deg} \times 1.46 \text{ kcal deg}^{-1} = 2.96 \text{ kcal}$$

Finally, we calculate that the oxidation of one mole (24.31 g) of magnesium will liberate

$$2.96 \text{ kcal} \times \frac{24.3 \text{ g}}{0.500 \text{ g}} = 144 \text{ kcal}$$

The heat capacity of a bomb calorimeter (including the water) is usually determined by an initial experiment in which a substance with known properties is ignited.

In the example just quoted, the heat capacity of the bomb and the surrounding water was first determined by igniting 0.400 g of solid benzoic acid. With this sample the temperature rise was found to be 1.73°C.

It is known, from very accurate work in other experiments, that the combustion of one mole (122 g) of benzoic acid releases 771 kcal of energy. We can use this information together with our calibration results to calculate how many calories of heat energy are required to raise the temperature of the calorimeter by 1°C.

If the ignition of 122 g of benzoic acid releases 771 of kcal of heat energy, then 0.400 g will release

$$0.400 \text{ g} \times \frac{771 \text{ kcal}}{122 \text{ g}} = 2.53 \text{ kcal}$$

This release of 2.53 kcal of heat energy produces a rise of 1.73°C in the bomb calorimeter. The heat capacity of the bomb must thus be

$$\frac{2.53 \text{ kcal}}{1.73 \text{ deg}} = 1.46 \text{ kcal deg}^{-1}$$

Our second example of the measurement of a heat change involves an endothermic process, the vaporization of water.

$$H_2O(l) \rightarrow H_2O(g) \quad (100°C, 1 \text{ atm}) \tag{8.2}$$

The heat change accompanying this reaction can be measured in the apparatus shown in Figure 8-2. Such an apparatus is easily constructed for undergraduate use.

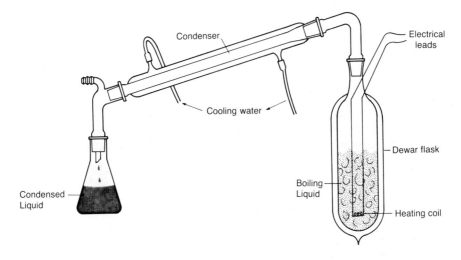

Figure 8-2. Apparatus for measuring heat of vaporization.

An electric current passes through a heating coil, causing the water inside an insulated Dewar flask to boil. The amount of water which distills out of the Dewar flask is measured by condensing it in a weighed flask.

In a typical experiment, a current of 0.350 amps was passed at 32.0 volts into the heating coil. The water which condensed over a period of five minutes was collected and found to be 1.38 g.

The heat energy generated electrically is given (in joules) by the product of the voltage, the current (in amperes) and the time (in seconds). We thus find that the total heat energy supplied to the system is given by:

$$\text{Heat energy} \quad = \quad \text{volts} \times \text{amps} \times \text{sec}$$

$$= \quad 32.0 \times 0.350 \times 300 = 3360 \text{ joules.}$$

Since 4.18 joules is equal to one calorie, we then have:

$$\text{Heat absorbed} = 3360 \text{ joules} \times \frac{1 \text{ cal}}{4.18 \text{ joules}} = 804 \text{ cal}$$

This 804 cal enables 1.38 g, or 0.0767 mole, of water to evaporate. The heat energy required to evaporate *one mole* of water at 100°C and 1 atmosphere pressure is thus

$$1 \text{ mole} \times \frac{804 \text{ cal}}{0.0767 \text{ mole}} = 10{,}500 \text{ cal}$$

$$= 10.5 \text{ kcal}$$

The accepted value for the molar heat of vaporization of water is 9.72 kcal mole^{-1}. Our measured value is too high because we have neglected to allow for heat losses in the evaporator. Not all the electrically produced heat energy goes toward evaporating the water; some of it escapes through the vacuum flask into the surroundings.

HEAT CHANGES AT CONSTANT VOLUME

The two processes we have just discussed differ from each other in that one is an exothermic and the other an endothermic change. There is also another, more subtle, difference between them. The first process takes place at constant volume, while the second takes place at constant pressure. In thermodynamics such a difference is of some importance.

Let us consider the constant volume process first. When magnesium is oxidized inside a bomb calorimeter, the volume of the reaction system is held constant by the strong container. Under such circumstances no work of any kind is done by the system. Only a heat change occurs. In particular, no work of expansion or contraction is involved since the volume remains unchanged.

Applying the first law,

$$\Delta E = q - w$$

to such a constant volume process, we must equate w to zero. This means that

$$q_v = \Delta E \qquad\qquad (8.3)$$

where the subscript v reminds us of the constant volume condition. The heat absorbed is thus exactly equal to the gain in internal energy.

This result is true for *any* change which occurs at constant volume, and not just for the oxidation of magnesium in a bomb calorimeter. In such a change no expansion work can be done, and if the system is arranged so as not to produce electrical work or any other work, the quantity w must be zero and equation (8.3) is valid. We can conclude quite generally that for any such process occurring at constant volume, *the heat absorbed is equal to the increase in the internal energy, ΔE.*

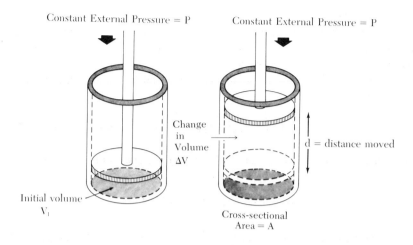

Figure 8-3. Cylinder and piston for demonstrating work of expansion.

WORK DUE TO EXPANSION AND CONTRACTION

Although no work is done during a constant volume process, the same is not true for a change occurring at *constant pressure*. At constant pressure the system will usually alter in volume, and work will be done either *by the system* as it expands and pushes back the atmosphere, or by the atmosphere *on the system* as it contracts. In Chapter 5, it was stated without proof that the magnitude of this work due to the expansion or contraction is given by the expression $P\Delta V$. Let us now establish this result.

Suppose that the system we are interested in is enclosed by a tightly fitting piston and cylinder such as that shown in Figure 8-3. As the constant pressure change occurs, so the volume of the system increases and the piston moves upwards against the external pressure P as shown in the figure. Since a force moves through a distance, energy is expended.

Suppose the piston has a cross-sectional area A and moves a distance d. At any point during the expansion, the total force acting on the cylinder will be the product of the external pressure times the area, that is, the quantity $P \times A$. Such a result follows from the definition of pressure as force per unit area.

In the course of the expansion, the force PA moves through a distance d. The work done as a result of this movement is $P \times A \times d$, the product of the force and the distance. Since the quantity Ad is equal to the increase in volume, we can write $Ad = \Delta V$, and hence $P \times A \times d = P\Delta V$. The expansion work done by the system as it vaporizes is thus the quantity $P\Delta V$.

This result holds for *any* expansion against constant pressure. We can write for *any* change at constant pressure in which only expansion work is done:

$$w = P\Delta V \tag{8.4}$$

Note that this equation is valid for changes in which the volume *decreases* as well as for changes in which it increases. If the volume decreases, ΔV and hence $P\Delta V$ are negative. Equation (8.4) tells us that w must then be negative. Physically, this means that work is done *on* the system by the atmosphere.

As an example of the use of equation (8.4), let us calculate the expansion work $P\Delta V$ for the reaction

$$2H_2(g) + O_2(g) \rightarrow 2H_2O(l) \quad (25°C, 1 \text{ atm}) \tag{8.5}$$

The volume of one mole of gas at $25°C$ $(298°K)$ and one atmosphere is given by:

$$22.4 \text{ liters} \times \frac{298 \text{ deg}}{273 \text{ deg}} = 24.5 \text{ liters}$$

The initial value of the volume in reaction (8.5) is thus $3 \times 24.5 = 73.5$ liters. In other words, $V_1 = 73.5$ liters. The final volume is that of two moles of H_2O; i.e., $V_2 = 36$ ml $= 0.036$ liters. Thus

$$\Delta V = V_2 - V_1 = -73.5 \text{ liters}$$

and

$$
\begin{aligned}
w = P\Delta V &= 1 \text{ atm} \times (-73.5) \text{ liters} \\
&= -73.5 \text{ liter atm} \\
&= -73.5 \text{ liter atm} \times \frac{24.2 \text{ cal}}{1 \text{ liter atm}} \\
&= -1.78 \text{ kcal}
\end{aligned}
$$

Another, more direct method of arriving at the same result is to neglect the volume of the liquid water, and use the gas equation

$$PV = nRT$$

The change in volume, ΔV, is then the volume of three moles of gas at this temperature and pressure; i.e.,

$$\Delta V = \frac{-3RT}{P}$$

where the negative sign indicates a contraction in volume. Thus:

$$w = P\Delta V = -3RT$$

Inserting values of R (in calories) and T, we find

$$w = -3 \text{ moles} \times 1.987 \text{ cal deg}^{-1} \text{ mole}^{-1} \times 298 \text{ deg}$$

$$= -1,780 \text{ cal} = -1.78 \text{ kcal}$$

which is the same value obtained previously.

In general, if the change in the number of moles of gas in a reaction is Δn, we can write

$$w = P\Delta V = (\Delta n)RT$$

Thus, for the reaction

$$CH_4(g) + O_2(g) \rightarrow CO_2(g) + 2H_2O(g) \quad (25°C, 1 \text{ atm})$$

$\Delta n = 1$ and

$$w = P\Delta V = RT$$

$$= 1 \text{ mole} \times 1.987 \text{ cal deg}^{-1} \text{ mole}^{-1} \times 298 \text{ deg}$$

$$= 592 \text{ cal}$$

while for the reaction

$$H_2(g) + Cl_2(g) \rightarrow 2HCl(g)$$

$\Delta n = 0$ and $w = P\Delta V = 0$.

HEAT CHANGES AT CONSTANT PRESSURE

Since a constant pressure process is accompanied by the occurrence of expansion work, we cannot equate the heat change to ΔE as in the constant volume case. The heat energy absorbed not only will cause particles to move from lower to higher energy levels, but will also provide the energy necessary to push back the atmosphere. For instance, when one mole of liquid water is vaporized at 100°C, the heat absorbed (9.72 kcal) does not all go to increasing the internal energy. As we saw in Chapter 5 (p. 96), ΔE is only equal to 8.98 kcal. The remaining 0.74 kcal is accounted for by the $P\Delta V$ expansion work.

In the general case, we can say that q_p, the heat absorbed by a constant pressure process, is equal to the change in internal energy ΔE plus the expansion work $P\Delta V$. In other words, we can write

$$q_p = \Delta E + P\Delta V \qquad \qquad (8.6)$$

for any process occurring at constant pressure.

Most of the reactions which we commonly encounter in the laboratory are constant pressure processes. In a typical laboratory situation we mix reagents together and allow them to react at atmospheric pressure, that is, at constant pressure. Any heat energy evolved or absorbed in such a reaction is the quantity q_p of equation (8.6), which depends not only on the change in internal energy but also on the expansion work $P\Delta V$. This means that if we want to find the value of ΔE for most laboratory reactions we must measure not only the heat change q_p, but the barometric pressure P and the change in volume ΔV as well.

THE ENTHALPY

The necessity of having to measure ΔV each time before we can obtain a value for ΔE is something of a nuisance. One way out of this difficulty is to define a new thermodynamic function which includes the expansion work. Such a function is called the *enthalpy* and is given the symbol H.

The enthalpy is defined by the equation

$$H = E + PV \qquad \qquad (8.7)$$

and has just the properties we require. We can easily see this by evaluating ΔH for a constant pressure change from an initial state designated by the subscript 1 to a final state designated by the subscript 2. The initial value of H is given by

$$H_1 = E_1 + PV_1 \qquad \qquad (8.8)$$

while the value of H for the final state (which is at the same pressure as the initial state) is given by

$$H_2 = E_2 + PV_2 \qquad \qquad (8.9)$$

Subtracting equation (8.8) from (8.9), we have

$$\Delta H = H_2 - H_1 = (E_2 - E_1) + P(V_2 - V_1)$$

or

$$\Delta H = \Delta E + P\Delta V \qquad \qquad (8.10)$$

Comparing this result with equation (8.6) leads us to the conclusion that

$$\Delta H = q_p \tag{8.11}$$

In other words, for any process occurring at constant pressure in which no work is done other than expansion work, *the heat energy absorbed is equal to the increase in enthalpy, ΔH.*

Such a result means that ΔH is relatively easy to measure. If we measure the heat change which takes place when a reaction occurs at constant pressure under such conditions that no work is done other than expansion work, then we have measured ΔH directly. There is no need to measure volume changes as would be necessary to obtain ΔE.

ENTHALPIES OF FORMATION

Because E, P, and V are all state functions, any combination of them will also be a state function. In particular, the combination

$$H = E + PV$$

will be a state function. Moreover, H is also an *extensive property*. If we double the quantity of matter in a system, the value of E will be doubled and so will the value of V. If P remains constant, H will also be doubled. Thus H is proportional to the quantity of matter in the system.

The fact that the enthalpy is both a state function and an extensive property means that we can add and subtract values of ΔH, using Hess's Law, in exactly the same way as we added and subtracted values of ΔE in Chapter 5. In order to do this, we need values of ΔH derived from calorimetric experiments similar to those described earlier in the chapter.

Over the years chemists have measured the values of the enthalpy changes for a very large number of reactions. To tabulate these data for every reaction that has been studied would take up several volumes of closely printed text. Fortunately, a great deal of space can be saved by using a method based on Hess's Law which makes it possible to have only one entry for each chemical compound. Such an entry is called the *standard enthalpy of formation* of the compound. The term *standard heat of formation* is also used.

The standard enthalpy of formation of a pure chemical substance is the enthalpy change, ΔH, which occurs when *one mole* of that substance is formed from its elements under the standard condition of one atmosphere pressure and the temperature specified. In this book we shall use the symbol ΔH_f^0 to indicate the standard enthalpy of formation of a compound. Unless otherwise stated, we shall assume that the temperature involved is $25°C$. A list of standard enthalpies of formation can be found in Appendix 3.

The definition of ΔH_f^0 given above necessarily implies that if the substance involved is *an element,* then its standard enthalpy of formation is *zero.* To form an element from itself at one atmosphere pressure and a given temperature involves no change in state and hence involves no change in enthalpy.

As an example of the use of such a table, let us calculate ΔH^0 for the reaction

$$2H_2O_2(l) \rightarrow 2H_2O(l) + O_2(g) \qquad\qquad (8.12)$$

at 25°C.

From Appendix 3 we find

Substance	ΔH_f^0
$H_2O_2(l)$	-44.9 kcal
$H_2O(l)$	-68.3 kcal
O_2	0.0 by definition

These values of ΔH_f^0 tell us what ΔH^0 is for the following reactions at 25°C.

Reaction	ΔH^0
$H_2(g) + O_2(g) \rightarrow H_2O_2(l)$	-44.9 kcal
$H_2(g) + \frac{1}{2}O_2(g) \rightarrow H_2O(l)$	-68.3 kcal

From the fact that H is an extensive property, it then follows that we can manipulate the enthalpies according to Hess's Law.

Reaction	ΔH^0
$2H_2O_2(l) \rightarrow 2H_2(g) + 2O_2(g)$	$-2(-44.9)$ kcal
$2H_2 + O_2 \rightarrow 2H_2O$	$2(-68.3)$ kcal
$2H_2O_2 \rightarrow 2H_2O + O_2$	$-2(-44.9) + 2(-68.3)$ kcal

Thus ΔH^0 for reaction (8.12) is -46.8 kcal.

If we look at the terms contributing to ΔH^0, we find that:

$$\Delta H^0 = 2\Delta H_f^0(H_2O) + \Delta H_f^0(O_2) - 2\Delta H_f^0(H_2O_2)$$

that is, we have added together the ΔH_f^0 values of the products and subtracted from this the ΔH_f^0 value of the reactants.

We can use this procedure for any chemical reaction. For a general chemical reaction of the form:

$$aA + bB + cC + \ldots \rightarrow mM + nN + \ldots \qquad (8.13)$$

we can write:

$$\Delta H^0 = \{ m\Delta H_f^0(M) + n\Delta H_f^0(N) + \ldots \}$$

$$- \{ a\Delta H_f^0(A) + b\Delta H_f^0(B) + c\Delta H_f^0(C) + \ldots \}$$

or more briefly:

$$\Delta H^0 = \Sigma \Delta H_f^0(\text{products}) - \Sigma \Delta H_f^0(\text{reactants}) \qquad (8.14)$$

The reason for the validity of equation (8.14) is easily seen if we divide the general reaction (8.13) into two successive hypothetical reactions. In the first reaction we imagine the reactants A, B, C, etc. decomposing into their constituent elements:

$$aA + bB + cC + \ldots \rightarrow \text{Elements} \quad (25^\circ C, 1 \text{ atm}) \qquad (I)$$

This process is the exact *reverse* of forming the reactants from their elements, so that the enthalpy change ΔH_I is given by

$$\Delta H_I = - \Sigma \Delta H_f^0(\text{reactants})$$

The second step is the formation of the eventual products out of the elements produced in Stage I. We can write this as:

$$\text{Elements} \rightarrow mM + nN + \ldots \quad (25^\circ C, 1 \text{ atm}) \qquad (II)$$

for which

$$\Delta H_{II} = \Sigma \Delta H_f^0(\text{products})$$

Now reaction (8.13) is the net result of reactions (I) and (II), so that

$$\Delta H^0 = \Delta H_I + \Delta H_{II}$$

$$= \Sigma \Delta H_f^0(\text{products}) - \Sigma \Delta H_f^0(\text{reactants})$$

which is the required result.

A further example makes the use of equation (8.14) clearer.
Example. Find ΔH^0 for the reaction

$$4NH_3(g) + 5O_2(g) \rightarrow 6H_2O(g) + 4NO(g) \qquad (8.15)$$

From Appendix 3 we find:

Substance	$\Delta H_f{}^0$ (kcal)	
$NH_3(g)$	-11.0	
$H_2O(g)$	-57.8	(Not $H_2O(l)$!)
$NO(g)$	$+21.6$	
O_2	0.0	

Using (8.14) we then obtain

$$\Delta H^0 = \Sigma\Delta H_f{}^0 \text{(products)} - \Sigma\Delta H_f{}^0 \text{(reactants)}$$

$$= [6(-57.8) + 4(21.6)] - [4(-11.0) + 5(0.0)]$$

$$= -260.4 + 44.0$$

$$= -216.4 \text{ kcal}$$

In using tables of standard enthalpies of formation one must be careful to note the physical state of the substance concerned. The example just worked involves $H_2O(g)$ and *not* $H_2O(l)$. Reference to Appendix 3 shows that $\Delta H_f{}^0$ for $H_2O(g)$ is -57.8 kcal, while that for $H_2O(l)$ is -68.3 kcal. It is the former value which is required.

IS ENTHALPY REALLY NECESSARY?

Having learned to manipulate the enthalpy, we will now turn to consider some of the difficulties which beginning students have with this property. In particular, they feel that the introduction of the enthalpy only adds an unnecessary complication. All of us have an intuitive feeling for what is meant by the internal energy, E. Both on the microscopic and on the macroscopic level it corresponds to all the energy there is in a system. Why, then, introduce an additional quantity H which is nearly, but not quite, the same thing as E?

The answer to such a question is that it is a matter of convenience. ΔH is ordinarily easier to measure than ΔE because it is exactly equal to the heat change in a constant pressure process. As a result, it has become a tradition among chemists to give the results of thermochemical experiments in terms of ΔH rather than ΔE. Although tables of standard enthalpies of formation, such as the one given in Appendix 3, are readily available, tables of standard internal energies of formation are rarely presented. Thus, even if you and I are prejudiced against the enthalpy, we have little option but to use it.

The difference between ΔE and ΔH for a chemical reaction is seldom a large quantity. In many cases the difference is immeasurably small. We know from equation (8.10) that

$$\Delta H = \Delta E + P\Delta V$$

and in most cases $P\Delta V$ is small compared to ΔE. Values of $P\Delta V$ and ΔE for several reactions have already been given in Table 5-1. Among the reactions shown in this table, the largest percentage difference between ΔE and ΔH is for the vaporization of water. For this reaction $P\Delta V$ is some 6 per cent of ΔE.

In Chapter 5 we learned how to use a table of mean bond energies to calculate an approximate value for ΔE. The error in such an approximation was sometimes larger than 5 kcal. The difference between ΔE and ΔH for a chemical reaction is seldom as large as 5 kcal. Accordingly, we can regard energy differences obtained from mean bond energies as being either approximate ΔE values or approximate ΔH values.

There is also very little point in distinguishing between ΔE and ΔH when using the low energy-high entropy rule to predict the direction in which a reaction will proceed. In practice ΔE always has the same sign as ΔH, so that the "low energy" side of a chemical equilibrium is also the "low enthalpy" side. In applying the rule, then, it makes no difference whether we say that the low energy side or the low enthalpy side is favored at low temperatures.

ENTHALPY TABLES AND EQUILIBRIUM

Previously we restricted our use of the low energy-high entropy rule mainly to gaseous reactions. It was easy to predict ΔE for such reactions from a table of bond energies. Now that we have access to tables of standard enthalpies of formation, such restrictions no longer apply. We can find ΔH^0 for any reaction provided only that sufficient ΔH_f^0 values are available. ΔS^0 values can also be deduced from tables. The number of reactions to which we can apply the low energy-high entropy rule is now very much broadened.

Consider, for instance, the reaction

$$2Ag(s) + I_2(s) \rightarrow 2AgI(s) \tag{8.16}$$

This is not a reaction whose direction we can predict by the methods employed previously. Because not all the bonding involved is covalent, we cannot use a table of bond energies to calculate an approximate value for ΔE^0 or ΔH^0. We can say very little about ΔS^0 for this reaction either, without consulting tables. Since only solids are involved in the reaction, ΔS^0 must be small, but it is not obvious whether it is positive or negative.

However, we can easily predict the behavior of reaction (8.16) by using the tables of measured thermodynamic data such as appear in Appendix 3. Using the ΔH_f^0 values from these tables we find that

$$\Delta H_{298}^0 = 2(-14.9) \text{ kcal}$$

$$= -29.8 \text{ kcal}$$

and

$$\Delta S_{298}^0 = 2(27.3) - \{ 2(10.2) + 27.8 \}$$

$$= +6.4 \text{ cal deg}^{-1}$$

Thus it is the right-hand side of reaction (8.16) which corresponds to both the lower enthalpy and the higher entropy at 25°C, and presumably at other temperatures as well. Use of the low energy–high entropy rule then leads to the prediction that silver is oxidized by solid iodine to silver iodide at all temperatures. Such a prediction agrees with experiment.

In Appendix 3, the ΔH_f^0 and ΔS^0 values for more than 300 substances are listed. These tables enable us to calculate ΔH^0 and ΔS^0 for a large number of reactions, the only restriction being that every substance featured in the reaction must also be listed in the table. Once we know ΔH^0 and ΔS^0 for a reaction, we can use the low energy–high entropy rule to predict the direction of the reactions at high and low temperatures.

In using thermodynamic tables in this way we must be sure to appreciate one important limitation on the information they can give. They can tell us in which direction equilibrium lies, but they cannot tell us *how long it will take* for the equilibrium to establish itself. In some cases equilibrium is established just as fast as the reactants are mixed; in other cases it takes longer than the lifetime of the universe. A real weakness of thermodynamics is that it cannot tell us at all when to expect either extreme.

The use of thermodynamic tables and of the low energy–high entropy rule often tells us that a certain reaction *can* occur, though experimentally there is no evidence for it ever having done so. Such a reaction is always one which occurs so slowly that we cannot notice in a finite period of time that any change has taken place.

As an example, consider the reaction

$$2NO_2(g) \rightarrow N_2(g) + 2O_2(g) \tag{8.17}$$

From Appendix 3 we find that ΔH^0 for this reaction is -15.8 kcal, while ΔS^0 is $+29.2$ cal deg^{-1}. Application of the low energy–high entropy rule would lead us to expect NO_2 to be an unstable compound, decomposing into its elements at any temperature. In fact, NO_2 is one of the best known oxides of nitrogen. It is a brown gas and shows no tendency to decompose into its elements at any easily attainable temperature. Reaction (8.17), although it is favored thermodynamically, occurs so slowly as to be undetectable.

Interestingly enough, NO_2 *does* decompose at moderately high temperatures, but *not* to nitrogen. It decomposes into another oxide, NO, according to the equation

$$2NO_2(g) \rightarrow 2NO(g) + O_2(g) \tag{8.18}$$

From tables, we find that ΔH^0 is +27.4 kcal, while ΔS^0 is +35.0 cal deg^{-1} for this reaction. We see that the sign of the entropy change favors reaction (8.18), while that of ΔH^0 tends to inhibit it. Accordingly, we expect the reverse of reaction (8.18) to occur at low temperatures, and the forward reaction to occur at high temperatures. Such behavior is observed experimentally. At room temperature NO reacts with oxygen to form NO_2. At higher temperatures, above about $500°C$, the reaction swings in the opposite direction and NO_2 decomposes into NO and O_2.

What have we learned from the two reactions we have just considered? We have learned that thermodynamics by itself does not answer every chemical problem. If we find from a table of thermodynamic data that a reaction *can* occur, this does not mean that it will take place in a reasonable time. The reaction may be one like reaction (8.17), which is too slow to be observable. We have also learned that if a given reaction is thermodynamically possible, this does not mean that all alternative reactions are forbidden. Reactions (8.17) and (8.18) are two alternative ways in which NO_2 can dissociate. At high temperatures thermodynamics tells us that both are possible, but only one occurs. NO_2 dissociates into NO and O_2 according to equation (8.18). The one reaction proceeds in preference to the other because it is faster. Which reaction is preferred is not a matter which thermodynamics can decide because it cannot tell us anything about the *rates* of reactions.

PROBLEMS 8

1. An electrical heating coil is immersed in a kilogram of water at $25°C$ and connected to a 20 V supply. A current of 1 amp passes for five minutes. What is the final temperature of the water?

2. Twenty-five liters of hydrogen at a total pressure of 1 atmosphere is produced by the action of an acid on a metal. Calculate the work done by the gas in pushing back the atmosphere in (a) liter atmospheres, (b) calories. (1 liter-atm = 24.2 cal)

3. Calculate the expansion work for the reactions listed below in calories or kilocalories from the relationship

$$P\Delta V = \Delta nRT$$

Neglect the volumes of solids and liquids.
   a. $2\,CO(g) + O_2(g) \rightarrow 2CO_2(g)$ $(25°C, 1\ atm)$
   b. $H_2O(g) + I_2(s) \rightarrow 2HI(g)$ $(25°C, 1\ atm)$
   c. $C_2H_5OH(l) \rightarrow C_2H_5OH(g)$ $(78.2°C, 1\ atm)$
   d. $H_2O(g) + Cu(s) \rightarrow CuO(s) + H_2(g)$ $(1000°K, 1\ atm)$
   e. $H_2O(g) + C(graphite) \rightarrow CO(g) + H_2(g)$ $(1000°K, 1\ atm)$

4. An electrical current of 0.368 amps at 32.0 volts was passed into the heating coil of the apparatus shown in Figure 8-2, and 3.71 g of ethanol, C_2H_5OH, was found to condense over a period of five minutes. Use this information to calculate ΔH^0 for the reaction

$$C_2H_5OH(l) \rightarrow C_2H_5OH(g)$$

at one atmosphere and the boiling point of ethanol $(78°C)$.

5. A sample of 0.550 g of naphthalene was burned in a bomb calorimeter in excess oxygen. The temperature rose from $23.862°C$ to $25.912°C$. In a separate experiment, the heat capacity of the calorimeter was found to be 2571 cal deg^{-1}. Calculate ΔE for the reaction

$$C_{10}H_{10}(s) + 12\tfrac{1}{2}O_2(g) \rightarrow 10CO_2(g) + 5H_2O(l)$$

under the conditions of the experiment.

6. 250 ml of 1M HCl and 250 ml of 1M Na_2SO_4, both at the same temperature, were mixed in a polystyrene container. The temperature of the resulting solution was found to have *fallen* by $2.6°C$. Assuming the heat capacity of the resulting solution to be 0.98 cal deg^{-1} for each ml, and neglecting heat losses through the polystyrene container, calculate ΔH for the reaction:

$$H_3O^+_{(aq)}\ (1M) + SO_4^{2-}(1M) \rightarrow H_2O(l) + HSO_4^-_{(aq)}\ (0.5M)$$

7. 1.89 g of solid benzoic acid, C_6H_5COOH, is placed in a bomb calorimeter and ignited in excess of oxygen. A temperature rise from $24.983°C$ to $25.615°C$ was observed. Find the heat capacity of the calorimeter, given that $\Delta E = 771.2$ kcal for the reaction

$$C_6H_5COOH(s) + 7\tfrac{1}{2}O_2(g) \rightarrow 7CO_2(g) + 3H_2O(l)$$

under the conditions of the experiment.

8. 3.20 g of methanol, CH_3OH, is placed in the same calorimeter used in the previous problem, and also ignited in excess of oxygen. A temperature rise from $24.802°C$ to $25.704°C$ was observed. Find ΔE for the reaction:

$$CH_3OH(l) + 1\frac{1}{2}O_2(g) \rightarrow CO_2(g) + 2H_2O(l)$$

at 25°C. From this, find ΔH for the reaction. Compare this result with that deduced from the thermodynamic tables in Appendix 3.

9. The following ΔH^0_{298} values can be determined from direct experimental measurements of the heat evolved when hydrogen gas, graphite, and liquid benzene (C_6H_6) are oxidized.

$2H_2(g) + O_2(g) \rightarrow 2H_2O(l)$ $\Delta H = -136.6$ kcal

$C(graphite) + O_2(g) \rightarrow CO_2(g)$ $\Delta H = -94.1$ kcal

$C_6H_6(l) + 7\frac{1}{2}O_2(g) \rightarrow 3H_2O(l) + 6CO_2(g)$ $\Delta H = -781.0$ kcal

Use these three values to calculate ΔH_f^0 for liquid benzene at 25°C.

10. The ΔH_f^0 value of red phosphorus is given in tables as -4.2 kcal. Why is the value not zero?

11. Use the table of standard enthalpies of formation given in Appendix 3 to calculate the value of ΔH^0 at 25°C for the following reactions.
    a. $S(rhombic) + O_2 \rightarrow SO_2(g)$
    b. $Fe(s) + S(rhombic) \rightarrow FeS(s)$
    c. $Hg(l) \rightarrow Hg(g)$
    d. $CS_2(l) \rightarrow CS_2(g)$
    e. $Mg(s) + \frac{1}{2}O_2(g) \rightarrow MgO(s)$
    f. $PCl_5(g) \rightarrow PCl_3(g) + Cl_2(g)$
    g. $2Al(s) + Fe_2O_3(s) \rightarrow 2Fe(s) + Al_2O_3(s)$
    h. $H_2(g) + I_2(s) \rightarrow 2HI(g)$
    i. $2H_2(g) + O_2(g) \rightarrow 2H_2O(l)$
    j. $4Li(s) + O_2(g) \rightarrow 2Li_2O(s)$

12. Using the table of ΔH_f^0 values given in Appendix 3, calculate ΔH^0 for the reactions given in Problem 5.6 (p.114). Compare these values to the approximate values for ΔE obtained previously using bond energies.

13. Use the tables of thermodynamic functions given in Appendix 3 to calculate ΔH^0 and ΔS^0 for the reactions listed below at 25°C. Then predict the direction of each reaction at very low and very high temperatures using the low energy–high entropy rule.
    a. $CH_4(g) + I_2(g) \rightleftharpoons CH_3I(g) + HI(g)$
    b. $Fe_2O_3(s) + 3CO(g) \rightleftharpoons 2Fe(s) + 3CO_2(g)$
    c. $N_2(g) + O_2(g) \rightleftharpoons 2NO(g)$
    d. $CO(g) + H_2O(g) \rightleftharpoons CO_2(g) + H_2(g)$
    e. $CdO(s) + H_2(g) \rightleftharpoons Cd(s) + H_2O(g)$
    f. $CuO(s) + H_2(g) \rightleftharpoons Cu(s) + H_2O(g)$
    g. $C(graphite) + H_2O(g) \rightleftharpoons CO(g) + H_2(g)$

14. For the process

$$Cl_2(l) \to Cl_2(g) \quad (240°K, 1 \text{ atm})$$

ΔH is found to be 4.87 kcal mole^{-1}. What is ΔE for this process?

15. Calculate both ΔH and ΔE for the following reactions at $25°C$ and 1 atmosphere pressure using tables.
   a. $H_2(g) + Br_2(l) \to 2HBr(g)$
   b. $H_2(g) + Br_2(g) \to 2HBr(g)$
   c. $H_2O_2(l) \to H_2O_2(g)$
   d. $H_2(g) + \frac{1}{2}O_2(g) \to H_2O(l)$

16. Calculate the mean bond energy for the C—H bond in CH_4 from the following enthalpy data:

Reaction	ΔH^0_{298}
$H_2(g) \to 2H(g)$	104.2 kcal
$C(graphite) \to C(g)$	171.3 kcal
$C(graphite) + 2H_2(g) \to CH_4(g)$	−17.9 kcal

17. Which of the enthalpy data given in the previous question is *not* available from Appendix 3?

18. Estimate the mean bond energy for the C—F bond in CF_4 from the enthalpy data given in Appendix 3.

19. Estimate the mean bond energy for the N—H bond in NH_3 from the enthalpy data given in Appendix 3.

Chapter 9

ENTROPY INCREASE AND FREE ENERGY DECREASE

THE LAW OF ENTROPY INCREASE

In order to make our approach to chemical equilibrium quantitative rather than qualitative, we must return to the point of view developed in Chapter 1. There we argued that an equilibrium state corresponds to a situation of maximum probability, and that the occurrence of a chemical reaction can be viewed as a movement from a *less probable* to a *more probable* state. It turns out that such a probabilistic view of equilibrium can be linked very easily to the property of entropy. How this can be done is best seen with the aid of an example.

The example we shall take is that of the flow of heat from a hotter body to a colder one. Consider the two systems, System a and System b, illustrated in Figure 9-1. Both systems consist of a mole of particles distributed over equally

	Level		Level	
	2 _____		_____ 2	
Figure 9-1. Two systems which are at different temperatures but are otherwise identical.	1	4.37×10^{10}	4.95×10^{10}	1
	0	6.023×10^{23}	6.023×10^{23}	0

System *a*	System *b*
(49.9°K)	(50.1°K)

Level					Level
	INITIAL STATE: STATE 1		FINAL STATE: STATE 2		
2					2
1	4.37×10^{10} $+$ 4.95×10^{10}	$\longrightarrow$	4.66×10^{10} $+$ 4.66×10^{10}		1
0	6.023×10^{23} 6.023×10^{23}		6.023×10^{23} 6.023×10^{23}		0
	System a System b		System a System b		

Figure 9-2. Heat flows from the hotter body to the colder one until both have the same distribution.

spaced energy levels. The energy levels are 3 kcal mole^{-1} apart. System a is at 49.9°K, while System b is at 50.1°K.

When these two systems are brought into contact, heat flows from the hotter to the colder body, until both systems have identical distributions, and hence identical temperatures (Figure 9-2).

The change illustrated in Figure 9-2 is from an improbable to the most probable state. Since the energy levels of Systems a and b are identical, one would expect, on a probability basis, that if energy is allowed to flow from one system to the other, both systems would end up with an equal share of energy. A situation in which System b has appreciably more energy quanta than System a is extremely unlikely, corresponding to one of those minuscule probabilities we encountered previously in dealing with out-of-equilibrium situations in Chapter 1.

Let us now look at this change in terms of thermodynamic probabilities. Let $W_{49.9}$ be the value of W_{max} for either system at 49.9°K, let $W_{50.0}$ be the value at 50.0°K, and let $W_{50.1}$ be the value at 50.1°K. Then W_1, the initial value of W_{max} for both systems considered together, is given by

$$W_1 = W_{49.9} \times W_{50.1}$$

(recall equation (6.1)). Similarly, the final value of W_{max}, which we shall call W_2, is given by

$$W_2 = W_{50.0} \times W_{50.0}$$

It is now possible to calculate W_2/W_1. This is an involved calculation and will not be attempted here. The result is:

$$W_2/W_1 = (10)^{10^{8.6}} \tag{9.1}$$

W_2 is thus enormously bigger than W_1, as expected.

We saw in Chapter 2 that, if only the two lowest energy levels of a system are occupied, the most probable distribution of a system has no serious competitors. Other distributions are improbable enough to be negligible. In effect, we can regard the most probable distribution as the *only* distribution of the system. In other words, the distribution shown in Figure 9-1 for System a at 49.9°K is the only distribution worth considering for this system at this temperature. The distribution shown for System b in the figure is likewise the only one which is at all probable at 50.1°K. Similar remarks will be true for both systems at 50°K. Such a state of affairs can only mean that the ratio W_2/W_1 of equation (9.1) represents the ratio of the probability of finding both moles at 50°K to that of finding one mole at 49.9°K and the other at 50.1°K. The flow of heat from a hotter to a colder body thus corresponds, like all processes which result in an equilibrium state, to a movement from a less probable to a more probable situation.

In one sense, our example has taught us nothing new. We were already aware that we can regard physical and chemical changes which occur of their own accord as movements towards a situation of maximum probability. Our advance has been not on a qualitative but on a *quantitative* level. We have succeeded in describing the movement towards equilibrium in terms of a *number,* a number which is, moreover, a *state function* of each of the systems involved. We were able to express the relative probabilities of the initial and the final states by the ratio W_2/W_1, where W_2 was the final value of W_{max} for both systems considered jointly, and W_1 was the initial joint value of W_{max} .

We will now take it as axiomatic that such a result is true not only for the change just considered but for *any* movement towards equilibrium. If W_1 is the value of W_{max} before a change occurs and W_2 the value of W_{max} when a final equilibrium state has been reached, then the ratio W_2/W_1 gives us the relative probability of the initial and final states, *no matter what the change may have been.*

For instance, if we were interested in the process

$$1 \text{ g Cu}(35°C, 1 \text{ atm}) + 1 \text{ g Cu}(15°C, 1 \text{ atm}) \rightarrow 1 \text{ g Cu}(25°C, 1 \text{ atm}) + 1 \text{ g Cu}(25°C, 1 \text{ atm})$$
$$\text{(9.2)}$$

for which

$$W_1 = \left\{ W_{max} \text{ for 1 g Cu at } 35°C \text{ and 1 atm} \times W_{max} \text{ for 1 g Cu at } 15°C \text{ and 1 atm} \right\}$$

and

$$W_2 = \left\{ W_{max} \text{ for 1 g Cu at } 25°C \text{ and 1 atm} \right\}^2$$

then the ratio W_2/W_1 will give us the relative probability of finding both grams of copper at a uniform 25°C as compared to finding the one gram at a uniform 35°C and the other at a uniform 15°C. Notice that W_2/W_1 does *more* than give the relative probability of finding each system *exactly* in its Boltzmann distribution at the appropriate temperature. When we talk about the probability of finding two grams of copper at 25°C we are not talking about the probability of finding them locked in their most probable distribution but rather of finding them fluctuating infinitesimally but incessantly over a "spread" of distributions around the Boltzmann distribution.

If our axiom that W_2/W_1 measures the relative probabilities of the final and initial states is always true, it leads us immediately to a very simple and completely general criterion for changes which result in an equilibrium state. In Chapter 1 we suggested that a final equilibrium state was *more probable* than the initial state. Reinterpreted in terms of W_{max}, this can only mean that W_2, the final value of W_{max}, must always be *larger* than W_1, the initial value. In mathematical symbolism:

$$W_2 > W_1 \qquad\qquad (9.3)$$

for any conceivable change which can occur in nature and which results in equilibrium.

The inequality (9.3) is usually expressed in terms of the entropy S rather than in terms of the thermodynamical probability W. S is not only easier to manipulate than W, but is also more easily related to heat changes and to the bulk behavior of a system. As we saw in Chapter 6, S is an increasing function of W_{max}; i.e., if W_{max} increases, so does S. This means that if $W_2 > W_1$, then $S_2 > S_1$. We can thus rewrite inequality (9.3) as:

$$S_2 > S_1 \qquad\qquad (9.4a)$$

or

$$\Delta S > 0 \qquad\qquad (9.4b)$$

Stated in words, either of these two inequalities means that WHEN A SYSTEM MOVES TO AN EQUILIBRIUM STATE THE ENTROPY INCREASES. Such a simple statement is one of the most important and widely useful laws in the whole field of science. It is called the SECOND LAW OF THERMODYNAMICS.

Unfortunately, we cannot explore all the implications of the Second Law in this book. They extend far beyond the boundaries of chemistry into physics and engineering on the one hand and into biology on the other. Here we shall confine our discussion of the second law to its applications to chemical equilibrium.

THE ENTROPY OF THE SURROUNDINGS

At first sight, a statement of the Second Law appears to contradict in a very fundamental way the view of chemical equilibrium so carefully built up in the preceding chapters. Previously, we argued that equilibrium depends on *two* factors, an energy change as well as an entropy change. Now we seem to be arguing that it depends only on *one* of these factors, namely an entropy change. How can we reconcile these views?

In order to answer this question, we must be careful to understand that the increase in entropy to which the Second Law refers includes *all* the entropy changes which accompany the move to equilibrium, and not just those that occur in the system itself. If you like, it is the entropy of the *universe* which increases as an equilibrium state is attained. Most of the changes in which we are interested in this book are chemical reactions, and such changes are invariably accompanied by the release or the absorption of heat energy. Not only do reactants turn into products, but the *surroundings are also altered* by the gain or loss of heat energy. An entropy change thus not only occurs in the reaction system but in the *surroundings as well*. The Second Law refers to the sum of *both* entropy changes.

As an example of how both reaction system and surroundings can change in entropy when a chemical reaction occurs, let us consider the oxidation of magnesium

$$2Mg(s) + O_2(g) \rightarrow 2MgO(s) \quad (25°C, 1 \text{ atm}) \tag{9.5}$$

in which both the initial and final states are at $25°C$ and 1 atmosphere pressure.

From our discussion of entropy in Chapter 6, we can predict that ΔS for reaction (9.5) is negative. The entropy of the two moles of solid on the right-hand side of reaction (9.5) must surely be less than that of the two moles of solid plus one mole of gas on the left-hand side. This conclusion is confirmed from the table of S^0 values in Appendix 3. We find that

$$\Delta S = \Delta S^0 = -51.22 \text{ cal deg}^{-1}$$

The fact that the reaction system has decreased in entropy does not necessarily mean that the Second Law has been violated. From Appendix 3 we find that

$$\Delta H = 2\Delta H_f^0 \text{ (MgO)} = -286 \text{ kcal}$$

The reaction is thus an exothermic one. When magnesium is oxidized, the surroundings have no option but to absorb heat and thus increase in entropy. We have only to show that this increase in entropy of the surroundings is large enough to offset the loss in entropy which occurs in the reaction system itself to be able to vindicate the Second Law.

For this purpose we shall assume an idealized set of surroundings. We assume that the reaction system is surrounded by a well-stirred liquid of such a good thermal conductivity and such a large heat capacity that, no matter how much heat is liberated as the reaction occurs, the temperature of the liquid is raised only infinitesmally above the original 25°C.

In Chapter 6 we saw how it was possible to link heat changes with entropy changes if the temperature remained constant and a Boltzmann distribution was maintained. The exact relationship is

$$\Delta S = \frac{q}{T}$$

where q is the amount of heat absorbed and T is the temperature. Applying this relationship to our idealized surroundings, we can write:

$$\Delta S_{surr} = \frac{q_{surr}}{T} \tag{9.6}$$

where ΔS_{surr} is the increase in the entropy of the surroundings, and q_{surr} is the heat energy *absorbed* by the surroundings.

Now the heat energy *absorbed* by the surroundings is equal to the heat *released* by the reaction system,

$$q_{surr} = {}^-q_{sys} \tag{9.7}$$

where the negative sign is necessitated by the convention that heat absorbed must be regarded as positive, while heat released is regarded as negative. Furthermore, since the chemical change occurring in the reaction system takes place at *constant pressure,* we can write equation (8.11),

$$q_{sys} = \Delta H$$

Substituting this result into (9.7) we obtain

$$q_{surr} = -\Delta H$$

so that

$$\Delta S_{surr} = \frac{q_{surr}}{T} = \frac{-\Delta H}{T} \tag{9.8}$$

(We can easily cross-check on the signs in equation (9.8). The oxidation of magnesium is exothermic so that ΔH is *negative.* If ΔH is negative, then (9.8) tells us that the surroundings increase in entropy. This agrees with the fact that they *absorb* heat energy.)

We can now evaluate the total entropy change, ΔS_{total}, of both the reaction system *and* the surroundings. It is given by:

$$\Delta S_{total} = \Delta S + \Delta S_{surr}$$

$$= \Delta S - \frac{\Delta H}{T} \qquad (9.9)$$

Now, by the Second Law the total entropy of system and surroundings taken together must increase as a result of the reaction, or

$$\Delta S_{total} > 0$$

so that by (9.9)

$$\Delta S - \frac{\Delta H}{T} > 0 \qquad (9.10)$$

Let us now check to see whether inequality (9.10) is valid for the oxidation of magnesium. Since $\Delta H = -286$ kcal and $\Delta S = -51.22$ cal deg^{-1} for the reaction, we have

$$\Delta S_{total} = \Delta S - \frac{\Delta H}{T}$$

$$= \left(-51.2 + \frac{286 \times 10^3}{298.2}\right) \text{cal deg}^{-1}$$

$$= (-51.22 + 959) \text{ cal deg}^{-1}$$

$$= 908 \text{ cal deg}^{-1}$$

Thus the total entropy increase is positive after all, and the Second Law is vindicated. Even though the reaction system itself loses entropy ($\Delta S = -51.22$ cal deg^{-1}), this loss is more than made up by the gain in entropy of the surroundings ($\Delta S_{surr} = +959$ cal deg^{-1}).

REACTIONS AT CONSTANT
PRESSURE AND TEMPERATURE

The arguments used above for the oxidation of magnesium at 25°C and 1 atmosphere pressure can obviously be applied in general. If *any* chemical or physical process occurs *at constant temperature and pressure* and ΔH and ΔS are the enthalpy and entropy changes of the reaction system, then the total increase in entropy is given by the equation

$$\Delta S_{tot} = \Delta S - \frac{\Delta H}{T}$$

for exactly the same reasons as apply in the case just discussed.

Application of the Second Law then yields the general result:

$$\Delta S - \frac{\Delta H}{T} > 0 \qquad \qquad \textbf{(9.10)}$$

for any change occurring at constant temperature and pressure which results in an equilibrium state.

The inequality (9.10) is usually employed in a slightly different form in chemical thermodynamics by multiplying through by $-T$. In many ways this is a pity, since, as it stands, inequality (9.10) points to the fact that it is the sum of *two* entropy changes which must be positive when a system moves to equilibrium, the entropy change of the system itself, ΔS, and the entropy change of the surroundings, $-\Delta H/T$.

If both sides of (9.10) are multiplied by T, we obtain the result

$$T\Delta S - \Delta H > 0$$

and further multiplication by -1 now leads to the usual form of the inequality:

$$\Delta H - T\Delta S < 0 \qquad \qquad \textbf{(9.11)}$$

The meaning of this inequality, stated in words, is this: if ΔH is the change in enthalpy and ΔS is the change in entropy of a reaction system when it moves to an equilibrium state, then the quantity $\Delta H - T\Delta S$ must be *negative*. Such a statement holds only for changes at constant pressure and at the constant temperature T.

The inequality (9.11) is a very important relationship. It is, in effect, a restatement of the low energy-high entropy rule in *quantitative terms*.

In order to see that this is the case, let us first consider how (9.11) behaves at low temperatures. If T is close to $0°K$ we can expect $T\Delta S$ to be small in comparison to ΔH. In the limit, we can neglect $T\Delta S$, and (9.11) then becomes the inequality

$$\Delta H < 0$$

This tells us that ΔH must be negative in a move to equilibrium. In other words, the low enthalpy side of a reaction will be favored at low temperatures. Neglecting the difference between ΔE and ΔH, this is then the same prediction that is given by the low energy-high entropy rule.

At high temperatures, by contrast, we can expect $T\Delta S$ to be a much larger quantity than ΔH. If T is high enough, we can neglect ΔH altogether. Inequality (9.11) then becomes

$$-T\Delta S < 0$$

Since T is always positive, this can only mean that

$$-\Delta S < 0$$

or, more simply, that

$$\Delta S > 0$$

A move to equilibrium at high temperatures thus corresponds to an increase in entropy of the reaction system. In other words, the high entropy side of a chemical equilibrium will be favored at high temperatures.

This new relationship described by (9.11) is, of course, more useful than the low energy-high entropy rule. In the first place, it is broader in scope. It applies not only to "high" and "low" temperatures but to intermediate temperatures as well. In the second place, it is a *quantitative* rather than a qualitative rule. Eventually it will enable us to predict exactly how much of what species is present at a given temperature in an equilibrium mixture.

THE FREE ENERGY

Before the implications of inequality (9.11) can be further explored, it is necessary to introduce a new thermodynamic function. This function is given the symbol G and is called the GIBBS FREE ENERGY or, more often, simply the FREE ENERGY. This new quantity is introduced purely for reasons of convenience. It facilitates calculations involving chemical equilibrium. However, it contributes nothing conceptual to our understanding of chemical equilibrium at a molecular level.

The free energy is defined in such a way that we can replace the inequality

$$\Delta H - T\Delta S < 0$$

by the much simpler inequality

$$\Delta G < 0$$

In order to do this, let us define the free energy by the equation

$$G = H - TS \qquad\qquad (9.12)$$

Consider how the quantity defined by equation (9.12) alters during a change at constant temperature and pressure from an initial state 1, to a final (equilibrium) state 2.

The initial value of the free energy, G_1, is given by

$$G_1 = H_1 - TS_1$$

and the final equilibrium value, G_2, by

$$G_2 = H_2 - TS_2$$

Subtracting these two equations we then have

$$(G_2 - G_1) = (H_2 - H_1) - T(S_2 - S_1)$$

or

$$\Delta G_{TP} = \Delta H - T\Delta S \qquad\qquad (9.13)$$

where the subscripts remind us of the conditions of constant temperature and pressure. We have already argued that if a system undergoes a change at constant temperature and pressure, then the following inequality holds:

$$\Delta H - T\Delta S < 0$$

Substituting this result into equation (9.13) gives the final desired inequality:

$$\Delta G_{TP} < 0 \qquad\qquad (9.14)$$

Stated in words, inequality (9.14) tells us that when a system moves to an equilibrium state, the free energy change is always *negative*. This can only mean that the equilibrium value of G is smaller than any non-equilibrium value. We thus arrive at the very important conclusion that AT CONSTANT TEMPERATURE AND PRESSURE, AN EQUILIBRIUM STATE CORRESPONDS TO A MINIMUM FREE ENERGY.

In Chapter 1, the hope was expressed that we would be able to discover what corresponds to the "downhill character" of a reaction. This task has now been accomplished, by introducing the free energy. In the same way that stones roll downhill to reach the lowest possible level, chemical reactions proceed until the lowest possible value of the free energy has been attained. Once such a minimum is reached, the reaction "stops;" i.e., the amount of each species present remains constant, and an equilibrium state results.

The way in which the free energy changes and achieves a minimum value as a reaction proceeds is best explained by means of simple examples. Before considering such examples, though, we need to know a few of the elementary properties of the free energy, G. Fortunately, these are easily disposed of.

We can argue that G, in terms of its definition

$$G = H - TS$$

is both a state function and an extensive property. G is a state function because it is a combination of three functions, H, T, and S, all of which singly are state functions. Similarly, G is an extensive property because both H and S are extensive properties. If G is both an extensive property and a state function, this means that we can add and subtract values of G and ΔG in exactly the same way that we can add and subtract values of the other thermodynamic functions such as H or S.

THE WATER-ICE EQUILIBRIUM

The first reaction we will consider is the conversion of one mole of ice into one mole of water at 25°C and 1 atmosphere pressure:

$$H_2O(s) \rightarrow H_2O(l) \quad (1 \text{ atm}, 25°C) \tag{9.15}$$

This is a particularly simple example, since no changes in concentration are involved.

As is well known, solid ice is unstable at 25°C and melts completely to liquid water. Let us see that this behavior is exactly what would be predicted from free energy considerations.

The free energy change for reaction (9.15) may be calculated from the following thermodynamic data:

	ΔH_f^0 (kcal)	S^0 (cal deg^{-1})
$H_2O(s)$	−69.984	10.63
$H_2O(l)$	−68.315	16.71

We find that $\Delta H = \Delta H° = -68.315 + 69.984 = 1.669$ kcal, while $\Delta S = \Delta S° = 16.71 - 10.63 = 6.08$ cal deg^{-1}. We can calculate a value for ΔG of

$$\Delta G = \Delta H - T\Delta S$$

$$= (1669 - 298.16 \times 6.08) \text{ cal}$$

$$= -144 \text{ cal}$$

In Figure 9-3, the line **AB** illustrates how the free energy of one mole of H_2O varies as it gradually becomes converted from ice to water at 25°C. In the figure the free energy of liquid water at 25°C is taken as zero. The free energy of one mole of ice is then 144 cal. As the number of moles of water increases, so

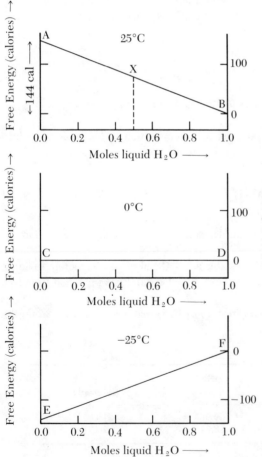

Figure 9-3. At 25°C, when one mole of H_2O is converted from ice to liquid, the free energy *decreases* by 144 cal (line AB). By contrast, the free energy *increases* for such a change at –25°C (line EF). At 0°C the free energy is unaltered on melting (line CD).

the free energy decreases regularly, so that AB is a straight line. That this must be so is seen if we consider the free energy when the reaction is half-way to completion at point X. At this stage we have half a mole of ice and half a mole of water. Since the free energy is an extensive property, the free energy at the half-way point is equal to the free energy of half a mole of ice plus the free energy of half a mole of water; i.e.

$$G_{0.5} = \frac{1}{2}G(\text{water}) + \frac{1}{2}G(\text{ice})$$

$$= \frac{1}{2}(0.0) + \frac{1}{2}(144)$$

$$= 72 \text{ cal}$$

The point X in Figure 9-3 is thus on a straight line between A and B. Because the free energy follows a straight line path from A to B in this way, the minimum value of G corresponds to the point B. The reaction goes to completion and *all* the ice turns into water.

Figure 9-3 also illustrates how the free energy varies as ice is converted into water at two other temperatures and 1 atmosphere. At $0°C$ there is no change of free energy as ice is converted into water, and any point along the line CD can be considered as a minimum free energy. In other words, we can mix ice and water together in any ratio at this temperature and still be at equilibrium.

The line EF in Figure 9-3 corresponds to the free energy change at $-25°C$. Here the minimum lies at E and the equilibrium state corresponds to pure ice. At this temperature, therefore, a mole of water will convert completely into a mole of ice.

THE $N_2O_4 \rightleftharpoons 2\ NO_2$ EQUILIBRIUM

Our next example of free energy change is the reaction

$$N_2O_4(g) \rightleftharpoons 2NO_2(g) \quad (25°C, 1 \text{ atm}) \tag{9.16}$$

The free energy difference between products and reactants for this reaction can be calculated from the following data:

	ΔH_f^0 (25°C) (kcal)	$\Delta S°$ (25°C) (cal deg^{-1})
$N_2O_4(g)$	2.19	72.70
$NO_2(g)$	7.93	57.35

We find that $\Delta H = \Delta H^\circ = 13.67$ kcal and $\Delta S = \Delta S^\circ = 42.00$ cal deg^{-1}. Consequently,

$$\Delta G = 13.67 - (42.00 \times 298.2) \times 10^{-3} \text{ kcal}$$

$$= +1.15 \text{ kcal}$$

Figure 9-4 shows how the free energy varies as one mole of N_2O_4 changes into two moles of NO_2 at 25°C and 1 atmosphere. The free energy is taken as zero for one mole of pure N_2O_4 at point A. Since ΔG for the reaction is $+1.15$ kcal, the free energy of two moles of NO_2 at point B is 1.15 kcal.

In this example the free energy exhibits a behavior quite different from that shown in the first example. Instead of increasing regularly along the straight line AXB as we might expect, the free energy first decreases to a minimum at the point Z and then increases through the point Y to the point B. The equilibrium position is thus at the point Z, where 0.814 moles of N_2O_4 are in equilibrium with 0.372 moles of NO_2.

In order to see why the free energy behaves in this fashion, let us consider

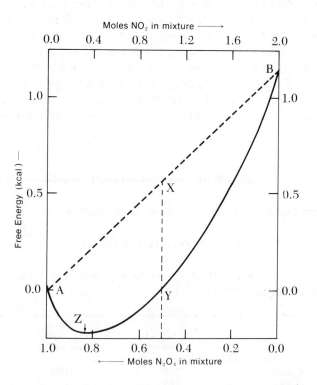

Figure 9-4. As one mole of N_2O_4 is converted into two moles of NO_2, the free energy follows the curve AZB. The point Z corresponds to equilibrium.

what happens when the reaction is half-way to completion. At this stage, half a mole of N_2O_4 has dissociated to give one mole of NO_2, leaving the other half mole of N_2O_4 unreacted. By analogy to the ice-water case just discussed, we might be tempted to write down the free energy for this half-way stage mixture as

$$G_{0.5} = \tfrac{1}{2}G^0 \, (N_2O_4) + G^0 \, (NO_2)$$

$$= 0.0 + 0.58 \text{ kcal}$$

$$= 0.58 \text{ kcal}$$

However, this value of G (which corresponds to the point X on the straight line AXB) is *incorrect*. The correct value of G for the half-way stage is lower and is given by the point Y. Why is this?

To answer this question we must refer to Figure 9-5. The free energy given by point X corresponds to state (a) of Figure 9-5, where the right amount of each gas is present but the two gases are in separate containers. Obviously, the total free energy in this case is equal to the sum of the individual free energies, 0.58 kcal.

Equally obviously, it is not State (a) but State (b) in the figure which corresponds to the half-way stage in the reaction. When a mole of N_2O_4 dissociates into two moles of NO_2, any intervening stage corresponds to a *mixture* of reactant and product rather than to the two gases in separate containers.

We can easily argue that the entropy of State (b) is higher than that of State (a). It is a matter of experience that if two different gases are given an opportunity to mix they will do so. This is because the mixture is a more probable state. It is extremely improbable that a mixture of gases will suddenly segregate into its pure components. Thus State (b) in Figure 9-5 is more probable than State (a) and has a higher entropy than it. The difference in entropy between the mixed gases is usually referred to as the *entropy of mixing*.

Now, an *increase* in entropy corresponds to a *decrease* in free energy (because of the *minus* TS term). The mixing of gases thus corresponds to a *lowering* of the free energy. This explains why the free energy of State (b) is less than that of State (a) in Figure 9-5, and why the point Y is below the point X in Figure 9-4. Similar remarks will apply at other stages in the decomposition of N_2O_4 to NO_2, and account for the fact that the free energy curve AZYB of Figure 9-4 is always *lower* than the straight line AXB.

We shall calculate the magnitude of the entropy of mixing and its quantitative effect on equilibrium in the next two chapters. For the moment we will confine our interest to the general qualitative effect such an entropy of mixing has on gaseous equilibria. If mixing had no effect on the free energy, then the

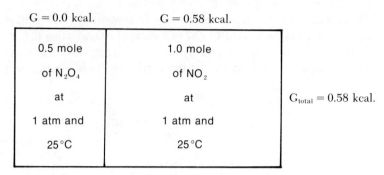

State (a)

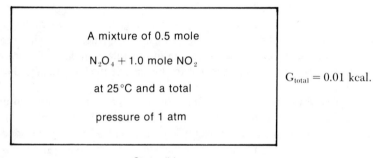

State (b)

Figure 9–5. State (a) represents the point X in Figure 9–4; it is less probable than State (b), the mixture of reactants and products, and thus has a *higher* free energy.

free energy for the decomposition of one mole of N_2O_4 into two moles of NO_2 would follow the straight line AXB of Figure 9–4, and the minimum free energy would correspond to pure N_2O_4 at A. *No* N_2O_4 would then decompose into NO_2. Instead of such behavior, we find that the effect of mixing is to lower the free energy curve so that it hangs down to a minimum, rather like a string held between the hands. As a result, some of the N_2O_4 is able to decompose into $2NO_2$, despite the fact that two moles of NO_2 have a higher free energy than one mole of N_2O_4.

It is instructive to look at the free energy behavior of reaction (9.16) at temperatures other than 25°C. Figure 9–6 illustrates how the free energy varies as one mole of N_2O_4 dissociates into two moles of NO_2 at one atmosphere pressure and a variety of temperatures ranging from 200°K to 450°K. At all temperatures the behavior is reminiscent of a string held between the hands. Instead of a straight line relationship, the free energy curve "sags" between the extreme values at the beginning and end of the reaction, because of the mixing effect discussed above. This "sag" always results in a minimum in the curve. The exact position of this minimum on each curve is not always obvious in the figure. Accordingly, it has been marked by an arrow in each case.

The position of this minimum varies almost as though it were a frictionless bead able to move along the string of our analogy. When one side of the reaction is very much higher in free energy than the other, the "bead" moves almost all the way over to the lower side. When reactant and product have very similar free energies, the bead stays somewhere near the center.

In more formal terms, Figure 9-6 demonstrates how the extent of this reaction depends on ΔG, the difference in free energy between pure reactants and pure products. Since all of our data refer to one atmosphere pressure, the value of ΔG is actually the standard value ΔG^0. Thus, if ΔG^0 for the reaction

$$N_2O_4(g) \rightarrow 2NO_2(g)$$

is large and *positive* (as occurs at low temperatures), very little reaction takes place, while if ΔG^0 is large and *negative* (as occurs at high temperatures) the reaction goes virtually to completion. Also, if ΔG^0 is small and both products and reactants have very similar free energies, then the equilibrium position corresponds to comparable quantities of both product and reactant.

In Table 9-1, the way in which the equilibrium position varies with the magnitude of ΔG^0 is illustrated. We notice that if ΔG^0 is more positive than +5 kcal, very little reaction occurs, while if ΔG^0 is more negative than -5 kcal, the reaction goes virtually to completion.

Table 9-1

Variation of Equilibrium Position of Reaction with the Value of
ΔG^0 for $N_2O_4 \rightarrow 2NO_2$

T ($^\circ$K)	Moles N_2O_4 at equilibrium	ΔG^0 (kcal)
200	0.999	+5.26
250	0.979	+3.17
275	0.928	+2.12
300	0.800	+1.07
325	0.500	+0.03
350	0.279	-1.02
400	0.038	-3.10
450	0.006	-5.18

There is a similar dependence of equilibrium position on the magnitude of ΔG^0 in all gaseous reactions, which we shall presently investigate. Before doing so, however, there are a few points relating to the decomposition of N_2O_4 which we must still clear up. In particular, we must explain why ΔG^0 is positive at

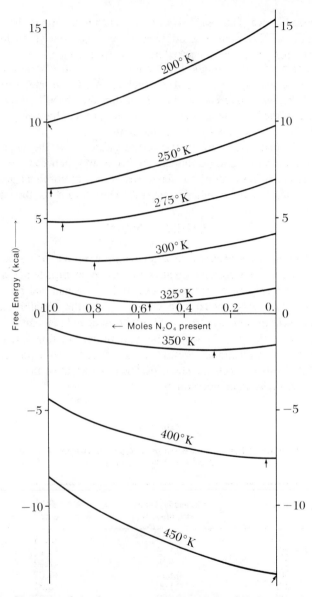

Figure 9-6. Variation of the free energy when one mole of N_2O_4 dissociates at various temperatures.

some temperatures and negative at other temperatures.

Since reaction (9.16) is a dissociation reaction, both ΔH^0 and ΔS^0 will be positive. Now, the free energy change ΔG^0 is given by the expression

$$\Delta G^0 = \Delta H^0 - T\Delta S^0$$

At low temperatures the $T\Delta S^0$ term will be small, and we can make the approximation

$$\Delta G^0 \approx \Delta H^0 \quad \text{(low T)}$$

Since ΔH^0 is positive, this means that ΔG^0 will be positive at low temperatures. At high temperatures, on the other hand, the $T\Delta S^0$ term will be very much larger than the ΔH^0 term. We can then make the approximation

$$\Delta G^0 \approx -T\Delta S^0 \quad \text{(high T)}$$

Since ΔS^0 is positive, this means that ΔG^0 will be *negative* at high temperatures.

We can calculate moderately accurate values of ΔG^0 for reaction (9.16) over a range of temperatures if we assume that both ΔH^0 and ΔS^0 do not vary with temperature. We have already found from tables that at $25°C$, $\Delta H^0 = 13.67$ kcal, while $\Delta S^0 = 42.00$ cal deg^{-1}. Using these values, we calculate that at $200°K$,

$$\Delta G^0 = \Delta H^0 - T\Delta S^0$$

$$= 13.67 - (200 \times 42.00) \times 10^{-3} \text{ kcal}$$

$$= 13.67 - 8.40 \text{ kcal}$$

$$= 5.27 \text{ kcal}$$

while at $400°K$,

$$\Delta G^0 = \Delta H^0 - T\Delta S^0$$

$$= 13.67 - (400 \times 42.00) \times 10^{-3} \text{ kcal}$$

$$= 13.67 - 16.80 \text{ kcal}$$

$$= -3.13 \text{ kcal}$$

The accepted values for ΔG^0 at these two temperatures, derived from experimental data, are $+5.26$ kcal at $200°K$ and -3.10 kcal at $400°K$, so that the agreement is remarkably good. In point of fact, such close agreement is somewhat fortuitous. At $1000°K$, the value of ΔG^0 calculated in this way is more than a kilocalorie in error.

DEPENDENCE OF EQUILIBRIUM POSITION ON THE
MAGNITUDE OF ΔG^0

Figure 9-7 illustrates the behavior of four gaseous reactions at 25°C and one atmosphere pressure. In this figure the free energy is plotted against the "extent" of the reaction. What is meant by this term is best illustrated by an example, say the reaction $N_2 + 3H_2 \rightarrow 2NH_3$. In this case, zero extent of reaction corresponds to one mole of N_2 and three of H_2, while an extent of unity corresponds to two moles of NH_3. A value of 0.5 will thus correspond to a mixture of ½ mole N_2 + 1½ mole H_2 + 1 mole NH_3.

We notice that the equilibrium position depends on the difference in free energy between reactants and products. The larger this difference, the closer the reaction approaches completion. The same point is made by Table 9-2, where ΔG^0 and the equilibrium value of the extent of each reaction are tabulated.

Both Table 9-2 and Figure 9-7 give us a feeling for what a free energy difference expressed in kilocalories means in terms of how far a reaction will go

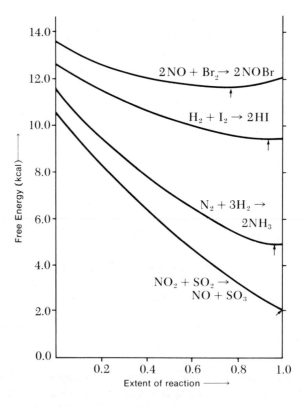

Figure 9-7. Free energy versus extent of reaction for four gaseous reactions at 25°C and 1 atmosphere total pressure. Arrows indicate equilibrium positions.

Table 9-2

Equilibrium Position and ΔG^0 for Various Gaseous
Reactions at 25°C and 1 Atmosphere.

Reaction	ΔG^0_{298} (kcal)	Extent of Reaction at Equilibrium
$2NO + Br_2 \rightarrow 2NOBr$	−1.98	0.667
$H_2 + I_2 \rightarrow 2HI$	−3.81	0.936
$N_2 + 3H_2 \rightarrow 2NH_3$	−7.88	0.969
$NO_2 + SO_2 \rightarrow NO + SO_3$	−8.51	0.999

in one direction or the other at 25°C. If ΔG^0 is zero or very small, comparable amounts of reactants and products will be present in the equilibrium mixture. A value of ΔG^0 as large as 10 kcal, though, is usually enough to swing the reaction almost completely to one side or the other. If ΔG^0 is more negative than −10 kcal, the reaction will be more than 99.9 per cent complete in most cases. Conversely, if ΔG^0 is more positive than 10 kcal, less than 0.1 per cent of the reactants can be changed into products.

Such a semiquantitative, rule-of-thumb approach can be useful in making an approximate assessment of a gaseous equilibrium. Consider, for instance, equilibrium (8.18),

$$2NO_2(g) \rightleftharpoons 2NO(g) + O_2(g)$$

which has already been discussed in terms of the low energy-high entropy rule in Chapter 8. There, we were able to predict that NO_2 decomposes at high temperatures, but not at low temperatures. But we were unable to go further than this, by making a prediction as to the behavior of NO_2 at a specific temperature, say 25°C. We can now improve on this situation.

We first calculate ΔG^0 for reaction (8.18) at 25°C. From Appendix 3 we find that $\Delta H^0 = +27.4$ kcal, while $\Delta S^0 = +35.0$ cal deg^{-1} at 25°C. Substituting these values into the expression for ΔG^0, we find that

$$\Delta G^0 = \Delta H^0 - T\Delta S^0$$

$$= (27.4 - 298 \times 35.0 \times 10^{-3}) \text{ kcal}$$

$$= 27.4 - 10.4 \text{ kcal}$$

$$= +17.0 \text{ kcal}$$

Since ΔG^0 is much larger than 10 kcal, we can conclude that at best only a small fraction of NO_2 will decompose into NO at 25°C.

Although this "±10 kcal" rule is useful, it must be emphasized that at best it is no more than a *rough guide*. In particular, it works best if the total number of molecules (reactants and products) is in the range from 4 to 7. If less than four molecules are involved, a quantity smaller than 10 kcal can be taken, while a larger quantity is indicated for more than 7 molecules. In the case just discussed the total number of molecules is 5, which is in the required range.

We should also realize that the rule only applies strictly to reactions occurring at 25°C and 1 atmosphere pressure. However, if the temperature and pressure are not very different from these standard conditions, the ±10 kcal limits can still be applied.

Another quantity which is useful in making an approximate assessment of the behavior of a gaseous equilibrium is the temperature at which $\Delta G^0 = 0$. At this temperature there will be comparable amounts of reactants and products present in the equilibrium mixture. Above this temperature one side of the reaction will be favored, while below it the other side will be favored.

This temperature can often be estimated with reasonable accuracy by assuming that the values of ΔH^0 and ΔS^0 found for 25°C from tables like those in Appendix 3 are also valid at other temperatures. As an example, let us take the reaction just considered, reaction (8.18). For this reaction

$$\Delta H^0_{298} = +27.4 \text{ kcal}$$

and

$$\Delta S^0_{298} = +35.0 \text{ cal deg}^{-1}$$

Assuming these values to hold at all temperatures, we can write

$$\Delta G^0 = \Delta H^0 - T\Delta S^0$$

$$= 27.4 \text{ kcal} - T(35.0 \times 10^{-3} \text{ kcal deg}^{-1})$$

The condition $\Delta G^0 = 0$ thus corresponds to the temperature

$$T = \frac{27.4 \text{ deg}}{35.0 \times 10^{-3}}$$

$$= 783°\text{K}$$

$$= 510°\text{C}$$

Below this temperature, NO_2 will decompose to a limited extent, if at all. In the vicinity of this temperature an appreciable percentage of NO_2 will decompose. At temperatures well above 510°C, NO_2 will dissociate almost completely.

STANDARD FREE ENERGIES OF FORMATION

We have already established that values of the free energy can be added and subtracted in the same way as the other thermodynamic functions, in accordance with Hess's Law. This fact allows us to use the same trick in the case of the free energy as was used to simplify the tabulation of values of ΔH. We can define and tabulate values of a quantity ΔG_f^0 in a manner analogous to the way we defined the quantity ΔH_f^0. This new quantity, ΔG_f^0, is usually called the STANDARD FREE ENERGY OF FORMATION.

The standard free energy of formation of a pure chemical substance is defined as the change in value of the free energy when *one* mole of the substance is formed from its elements. Both the elements and the substance must be at the same temperature and at the standard pressure of one atmosphere. Tables of ΔG_f^0 values are usually given at 25°C. Such a table can be found in Appendix 3.

If tables of ΔH_f^0 and S^0 are available, the values of ΔG_f^0 can be calculated from these other tables, but it is nevertheless useful to have the calculation already performed for us. As an example of such a calculation, let us compute ΔG_f^0 for NH_3 gas. We require ΔG for the reaction

$$\tfrac{1}{2}\,N_2(g) + 1\tfrac{1}{2}\,H_2(g) \rightarrow NH_3(g) \quad (25°C, 1 \text{ atm})$$

We can find ΔH for this reaction from enthalpy tables; it is simply ΔH_f^0 for $NH_3(g)$. From Appendix 3 we find that $\Delta H_f^0(NH_3) = -11.0$ kcal.

The value of ΔS for this reaction can be found from the S^0 values for the three gases involved. It is $46.0 - 1/2(45.8) - 3/2(31.2) = -23.7$ cal deg^{-1}. Thus,

$$\Delta G = \Delta H - T\Delta S$$

$$= -11.0 + (298.2 \times 23.7) \times 10^{-3} \text{ kcal}$$

$$= -11.0 + 7.1 = -3.9 \text{ kcal}$$

The value we have just calculated, namely -3.9 kcal, is also the figure given in Appendix 3 as the ΔG_f^0 value for NH_3.

Tables of standard free energies of formation are handled very much like tables of standard enthalpies of formation, and enable us to calculate the value of ΔG for any reaction at 25°C and 1 atmosphere pressure, provided that all the reactants and products appear in the table.

As an example, let us calculate ΔG^0 for the reaction

$$CO(g) + NO_2(g) \rightarrow CO_2(g) + NO(g) \quad (25°C, 1 \text{ atm}) \tag{9.17}$$

We find, using Appendix 3, that

$$\Delta G^0 = \Delta G_f^0(CO_2) + \Delta G_f^0(NO) - \Delta G_f^0(CO) - \Delta G_f^0(NO_2)$$

$$= -94.3 + 20.7 + 32.8 + 12.3$$

$$= -28.5 \text{ kcal}$$

Since ΔG^0 is much more negative than -10 kcal, this result means that reaction (9.17) goes to completion at $25°C$.

The reader should note the exact significance of the free energy ΔG^0 which has just been calculated. It refers to pure samples of all reactants and products. The initial free energy referred to is that of a mole of CO at $25°C$ and 1 atmosphere plus that of a mole of NO_2 at the same temperature and pressure *unmixed and in separate containers.* The final free energy referred to is again *not* a mixture of the products but *separate quantities* of one mole of CO_2 and one mole of NO at $25°C$ and 1 atmosphere pressure. Although the ΔG^0 calculated from a table of ΔG_f^0 values in this way often refers to the free energy change of a rather hypothetical reaction, it is a very useful quantity all the same.

WHAT IS FREE ENERGY?

We have now become sufficiently familiar with free energy to see that it is the quantity to use in dealing with chemical equilibrium and with the direction of chemical reactions. Even though we still have some way to go in predicting the exact position of equilibrium at any temperature, we are nevertheless able to make approximate predictions. For a given temperature we can now tell whether a gaseous reaction will occur to a negligible extent, or whether it will proceed to a limited extent, or whether it will go virtually to completion.

Even though we have been making real advances in this direction, we seem to have gotten further and further away from what is after all the very life blood of chemistry, namely the behavior of molecules. When we dealt with equilibrium in terms of enthalpy and entropy, we were dealing with properties to which we could attach some kind of molecular aspect. Is it possible to attach an equivalent microscopic significance to the free energy? The answer is no!

The real point about free energy is that it is a *manipulative convenience.* We have already seen in previous chapters what molecular features are important in determining chemical equilibrium, and we can translate these molecular features into the enthalpy, the entropy and the temperature. By introducing G, we are adding *nothing* to these qualitative insights, but are merely making the most convenient combination of S, H, and T. The introduction of G makes it very easy to handle chemical equilibrium, but it introduces no *new* factor not previously encountered. To try and identify the free energy with some obvious feature of molecular behavior is thus a waste of time.

There is another difficulty about the free energy – its name. Although G has the dimensions of energy, it is not really an energy, like kinetic energy or potential energy. In particular, it is not conserved. Although the energy of the universe remains constant, the free energy of the universe is getting less and less with time. Furthermore, this continual loss of free energy is not balanced out by a gain in other forms of energy elsewhere.

PROBLEMS

1. Draw a diagram, similar to Figure 9-3, showing how the free energy changes as one mole of white phosphorus is converted into one mole of red phosphorus at 1 atmosphere pressure and 25°C. (Obtain free energy data from Appendix 3.)

2. Draw a diagram, similar to Figure 9-3, showing how the free energy changes as two moles of solid Ag and one mole of solid I_2 react to form two moles of solid AgI according to the equation

$$2Ag(s) + I_2(s) \rightarrow 2AgI(s) \quad (25°C, 1 \text{ atm})$$

3. For the following process,

$$2H_2(g) + O_2(g) \rightarrow 2H_2O(l) \quad (25°C, 1 \text{ atm})$$

calculate from tables
   a. The change in entropy of the reaction system.
   b. The change in entropy of the surroundings.
   c. The total change in entropy.

4. When one mole of N_2O_4 gas is converted into two moles of NO_2 gas at 350°K and 1 atmosphere total pressure, the free energy varies with extent of reaction as follows.

Extent	G (kcal)
0.00	0.000
0.10	-0.465
0.20	-0.735
0.30	-0.930
0.40	-1.073
0.50	-1.174
0.60	-1.237
0.70	-1.264
0.80	-1.252
0.90	-1.190
1.00	-1.019

Plot these results on a graph and find the position of the minimum. Compare your answer to that given in Table 9-1.

5. Use the table of ΔG_f^0 values in Appendix 3 to find the values of ΔG^0 at 25°C for the following reactions. In each case predict whether the reaction will go virtually to completion, scarcely occur at all, or occur only to a limited extent, at 25°C.

  a. $CH_4(g) + 2O_2(g) \rightarrow CO_2(g) + 2H_2O(g)$
  b. $2F_2(g) + O_2(g) \rightarrow 2F_2O(g)$
  c. $2Cl_2(g) + O_2(g) \rightarrow 2Cl_2O(g)$
  d. $I_2(g) + 2HBr(g) \rightarrow 2HI(g) + Br_2(g)$
  e. $CO(g) + Cl_2(g) \rightarrow COCl_2(g)$
  f. $2CO(g) + O_2(g) \rightarrow 2CO_2(g)$
  g. $PCl_3(g) + Cl_2(g) \rightarrow PCl_5(g)$
  h. $N_2(g) + 3H_2(g) \rightarrow 2NH_3(g)$
  i. $CH_4(g) + I_2(g) \rightarrow CH_3I(g) + HI(g)$
  j. $CH_4(g) + 2H_2S(g) \rightarrow CS_2(g) + 4H_2(g)$
  k. $2NO(g) + Cl_2(g) \rightarrow 2NOCl(g)$
  l. $2NO(g) + Br_2(g) \rightarrow 2NOBr(g)$

6. Calculate both ΔH^0 and $T\Delta S^0$ at 25°C for the reactions given in the previous problem. Determine for each reaction whether ΔH^0 or $T\Delta S^0$ is numerically larger and hence which of these two factors has the predominating influence on the sign and magnitude of ΔG^0 at 25°C. In other words, find which of these reactions are "entropy determined" and which are "enthalpy determined" at this temperature.

7. Show that when a change occurs at *constant volume* and constant temperature, the entropy of the surroundings is increased by an amount

$$\Delta S_{surr} = \frac{-\Delta E}{T}$$

Hence show that if such a change results in an equilibrium state, then

$$\Delta E - T\Delta S < 0$$

8. For the process

$$CH_3COOH(s) \rightarrow CH_3COOH(l) \quad (25°C, 1 \text{ atm})$$

ΔH^0 is found to be +2.8 kcal while ΔS^0 has the value +9.7 cal deg^{-1}. Use these data to estimate the melting point of acetic acid. (Observed melting point = 17°C.)

9. For the process

$$A(s) \rightarrow A(l) \quad (25°C, 1 \text{ atm})$$

$\Delta H^0 = +1.67$ kcal and $\Delta S^0 = +6.08$ cal deg^{-1}. Find the melting point of A, and hence identify it.

10. Estimate from the data given in Appendix 3 at what temperatures white and red phosphorus can co-exist in equilibrium.

11. The following equilibria have all been discussed in Chapter 7. Estimate at what temperature we can expect the products to be favored over the reactants at 1 atmosphere pressure.

   a. cyclopentane $\rightleftharpoons$ 1-pentene

   b. $4HCl(g) + O_2(g) \rightleftharpoons 2H_2O(g) + Cl_2(g)$

   c. $N_2(g) + 3H_2(g) \rightleftharpoons 2NH_3(g)$

   d. $N_2(g) + 3Cl_2(g) \rightleftharpoons 2NCl_3(g)$

   e. ethylene $\rightleftharpoons$ polyethylene

   f. $HN{=}NH(g) \rightleftharpoons N_2(g) + H_2(g)$

12. Calculate ΔH^0 and ΔS^0 for the water gas reaction

$$C(graphite) + H_2O(g) \rightarrow CO(g) + H_2(g)$$

at $25°C$. Assuming these values to apply at all temperatures, estimate the temperature above which we can expect CO and H_2O to be formed in appreciable quantities from graphite and steam.

13. Estimate ΔH^0 and ΔS^0 for the dissociation of the following gases into their atoms: CO, F_2, H_2, HI, I_2, O_2. Then calculate the approximate temperature at which these gases will be half dissociated, i.e., the temperature at which $\Delta G^0 \approx 0$.

14. a. The standard molar entropy at $25°C$ of the very reactive radical CH_3 has been estimated as 46.4 cal deg^{-1}. Use this value and a table of bond energies to calculate ΔG^0_{298} for the reaction:

$$H_3C{-}CH_3 \ (g) \rightarrow 2CH_3 \ (g)$$

Next make a rough estimate of the temperature above which ethane (C_2H_6) is more than half dissociated into methyl radicals.

   b. Is the valence of carbon still 4 above this temperature?

15. If all chemical equilibria were governed by ΔS^0 rather than by ΔG^0, would the concept of valence have any meaning?

16. In a gaseous reaction of the type

$$A(g) + 2B(g) \rightarrow 2C(g)$$

ΔS^0 is usually of the order of -40 cal deg^{-1}. How negative must ΔH^0 be for such a reaction if it is to go to completion at $25°C$?

17. Discuss this statement: "All gaseous reactions in which more bonds are made than are broken will go to completion at $25°C$."

18. Compounds containing carbon, hydrogen, and one chlorine atom tend to be unstable at high temperatures because they decompose into HCl and a molecule containing a carbon—carbon double bond. The simplest of these reactions is the decomposition of ethyl chloride,

C_2H_5Cl:

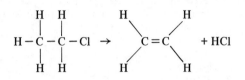

Make a rough estimate of the temperature at which this process will occur to an appreciable extent. Use the tables in Appendix 3.

Chapter 10

IDEAL
GASES

Before we can go any further with our study of chemical equilibrium, we must investigate the thermodynamic properties of ideal gases. We will not be able to predict where the free energy minimum corresponding to equilibrium occurs until we have some formula for predicting the free energy changes which occur when gases mix. In order to arrive at such a formula we must first investigate how the enthalpy and entropy of a gas change as it expands or contracts at constant temperature, and also what changes in these properties occur when different gases mix without reacting chemically.

ENTROPY AND ENTHALPY CHANGES ON EXPANSION

Consider the situation illustrated in Figure 10-1. One mole of xenon, assumed to behave as an ideal gas, is in container A. This container is separated from container B by a closed tap. Container B is completely evacuated.

When the tap is opened, gas expands from A into B until an equilibrium situation is obtained in which both containers are filled with xenon gas. The pressure and the temperature of the gas will then be uniform throughout each container. Moreover, the pressure and temperature in container A will be the same as the temperature and pressure in container B. Such a state of affairs is indicated diagrammatically in Figure 10-2.

A   B

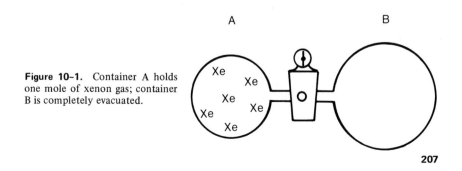

Figure 10-1. Container A holds one mole of xenon gas; container B is completely evacuated.

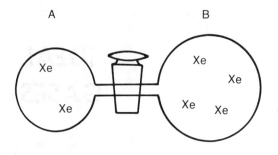

Figure 10-2. When the tap is opened, the xenon expands into container B until the pressure and temperature in both containers are the same.

In the discussion which follows, we shall refer to the initial state of the gas, depicted in Figure 10-1, as State 1; the final equilibrium state of the gas, depicted in Figure 10-2, we shall call State 2.

Let us first argue what the change in internal energy of the xenon gas will be as a result of this expansion. As the xenon expands to fill B, the molecules become, on the average, further away from each other. In real gases, the molecules attract or repel each other slightly. An increase in the average inter-molecular distance thus results in either an increase or a decrease in the total potential energy of the gas molecules. If the expansion of the gas is so rapid that no heat can be given to or taken from the surroundings, then the molecules have no way of increasing or decreasing their total energy. Thus, if they gain in potential energy by being separated from each other, they must balance out this gain by the loss of an exactly equivalent amount of kinetic energy in order to keep the total of potential and kinetic energy constant. We conclude that if the molecules of a gas attract each other so that the separation of molecules causes an *increase* in potential energy, the sudden expansion of the gas from the situation depicted in Figure 10-1 to that depicted in Figure 10-2 will result in a *cooling* of the gas due to the loss of kinetic energy by the molecules. Conversely, if the molecules of a gas *repel* each other, an expansion such as the one just described will result in a *loss* of potential energy as the molecules separate. This loss will be balanced by a gain in kinetic energy with a resultant *increase* in the temperature of the gas.

Intermolecular forces in a gas, whether attractive or repulsive, are in general small and only affect the properties of a gas to a small degree. We should thus expect the heating and cooling effects deduced above to be small. This expectation is confirmed experimentally. The effects are so small that they are rather difficult to measure. In fact, the first attempts to measure them were unsuccessful.

In the case of an ideal gas, such as we have supposed xenon to be, there are, by definition, no attractive or repulsive forces. When an ideal gas expands, we thus expect no heating or cooling effect, however slight. The molecules move, on average, further away from each other without any change in potential or kinetic energy.

Despite the above argument, many students may still have a sneaking suspicion that the writer has made an error somewhere. It seems contrary to experience that a gas can move from the state depicted in Figure 10-1 to that depicted in Figure 10-2 without a noticeable decline in temperature. Perhaps this is because all the experiences we have had of gases expanding have been ones in which the gas cooled. A closer inspection of such cases, for instance air escaping suddenly from a tire, reveals that the gas is not *only* expanding but also *pushing back the atmosphere.* If the gas pushes back the atmosphere it is doing *external work.* To do this work the gas must obtain the energy from somewhere. The energy comes from the kinetic energy of the molecules. This decrease in kinetic energy corresponds, of course, to a decrease in temperature.

Thus when an ideal gas not only expands but *does work while expanding,* it cools down. When it expands *without* doing external work of any kind it *will not cool down.* When xenon expands into a vacuum, as described above, *no external work is done.* No force moves through any distance. The gas will not cool down, nor will it heat up. No heat energy will be exchanged with the surroundings. In consequence we may conclude that in the equation describing the first law, namely:

$$\Delta E = q - w$$

both q and w are zero, and hence so is ΔE. In other words, there is no change in the internal energy of a gas when it expands into a vacuum. On a molecular level, when an ideal gas expands, neither the potential energy nor the kinetic energy of the molecules is altered. The internal energy of the gas, which is the sum of these two quantities, thus remains unaltered.

When an ideal gas expands from State 1 of Figure 10-1 to State 2 of Figure 10-2, the temperature remains unaltered. By Boyle's Law, therefore, $P_1 V_1 = P_2 V_2$, and the product PV remains constant; i.e., $\Delta(PV) = 0$. For the change under consideration

$$\Delta H = \Delta(E + PV)$$

$$= \Delta E + \Delta(PV)$$

$$= 0 + 0$$

There is thus no change in the enthalpy. We can conclude that when an ideal gas expands from any volume V_1 to any larger volume V_2 *at the same temperature, neither the internal energy nor the enthalpy will be altered.*

The same, however, cannot be said of the entropy. When the tap in Figure 10-1 is opened, the gas is no longer at equilibrium. It attains a new equilibrium state by expanding into container B, as shown in Figure 10-2. By the second law, this movement to equilibrium must be accompanied by an *increase in entropy.* There is no exchange of heat energy with the surroundings; in fact, the surroundings suffer no change at all as a result of the expansion. All the entropy increase must occur in the gas itself. The entropy of the gas in State 2 must thus be larger than the entropy of the gas in State 1. But by how much? In order to find ΔS we must calculate the relative probabilities of State 1 and State 2.

Let us assume that the volume of container A is 10 liters while that of container B is 20 liters. Thus V_1 = 10 liters and V_2 = 20 + 10 = 30 liters. In State 1, all the molecules are to be found in container A, while in State 2, they are to be found in both containers. Let us concentrate first on one individual molecule. What is the probability, in State 2 of its being in container A? Assuming that the molecule shows no preference for any particular container or any particular part of either container, the probability of finding the molecule in a particular volume is proportional to that volume. Thus the probability of finding the molecule in container A is 10/30 or 1/3.

Next, let us consider the probability of two molecules both being in A. Whether or not one molecule is in container A has no effect on the probability of the second being there. The location of the molecules, and their probabilities, are independent of each other. We can thus apply the law of multiplication of independent probabilities. The probability of finding two particular molecules in container A, is then $1/3 \times 1/3 = (1/3)^2$.

Similarly, the probability of finding *three* given molecules all in A is $(1/3)^3$, and so on for larger numbers of molecules. Finally, the probability of finding a *mole* of molecules *all* in container A is obviously $(1/3)^N$, where N is the Avogadro number. This number, $(1/3)^N$, is the ratio of the probability of State 1, in which all the molecules are in container A, to the probability of State 2, in which molecules occupy both containers. It thus represents the probability of the change

$$Xe\ (V = 10\ \text{liters}) \rightarrow Xe\ (V = 30\ \text{liters}) \qquad \textbf{(10.1)}$$

reversing itself.

We expect such a probability to be an unimaginably small number, so that we are not at all surprised to find that:

$$\left(\frac{1}{3}\right)^N = \frac{1}{3^{(6 \times 10^{23})}} = \frac{1}{10^{(3 \times 10^{23})}}$$

An alternative way of looking at the relative probabilities is in terms of energy levels. If we knew enough about translational energy levels, we could work out the thermodynamic probability of the Boltzmann distribution corresponding to State 1, namely W_1, as well as W_2, the equivalent thermodynamic probability of State 2. As we saw in Chapter 6, the ratio of these two quantities, W_1/W_2, gives us the relative probability of State 1 as opposed to State 2.

Such a quantum treatment is beyond the scope of this book; but no matter.

We have already calculated W_1/W_2 by using a classical rather than a quantum model, since we have already calculated the *relative probability* of State 1 with respect to State 2. It is $(1/3)^N$. We may therefore write:

$$\frac{W_1}{W_2} = \left(\frac{1}{3}\right)^N \text{ or } \frac{W_2}{W_1} = 3^N$$

Recalling that $S = k \log_e W$, we have:

$$\Delta S = S_2 - S_1 = k \log_e W_2 - k \log_e W_2 = k \log_e \frac{W_2}{W_1}$$

$$= k \log_e(3^N) = Nk \log_e(3)$$

$$= R \log_e(3)$$

In the usual units,

$$\Delta S = 2.18 \text{ cal deg}^{-1}$$

(Notice that what we normally regard as quite a small change in the entropy corresponds to a gargantuan change in probabilities.)

By an identical argument, we can calculate the changy in entropy for the general case, when *one mole* of an ideal gas expands from an initial volume V_1 to a final volume V_2. In such a case we have:

$$\frac{W_1}{W_2} = \left(\frac{V_1}{V_2}\right)^N$$

and hence

$$\Delta S = k \log_e \left(\frac{W_2}{W_1}\right) = k \log_e \left(\frac{V_2}{V_1}\right)^N$$

or finally

$$\Delta S = R \log_e \frac{V_2}{V_1} \qquad \textbf{(10.2)}$$

Observe that the entropy of a gas *increases* as its *volume increases*. By equation (10.2), if the final volume V_2 is greater than the initial volume V_1, then ΔS is positive. This is precisely what we would expect from our discussion in Chapter 6. In a larger volume, the constraint on the translational motion of the molecules is smaller. This means that the translational energy levels will be closer to each other, and that the translational entropy and hence the total entropy will be larger.

Instead of being expressed in terms of the initial and final volumes as in (10.2), the change of entropy of a gas can be expressed in terms of the initial and final pressures. Since the temperature remains constant, Boyle's Law holds. Thus:

$$P_1 V_1 = P_2 V_2$$

or

$$\frac{V_2}{V_1} = \frac{P_1}{P_2}$$

Substituting this in (10.2), we obtain

$$\Delta S = R \log_e \frac{P_1}{P_2} \tag{10.3}$$

Equations (10.2) and (10.3) refer to the entropy change of one mole of gas. If n moles of gas are involved, the entropy change will be n times larger than in the case of one mole. (Entropy being an extensive property, both the initial entropy, S_1, and the final entropy, S_2, will be proportional to the number of moles, n.) The formulae for n moles of gas, corresponding to equations (10.2) and (10.3), are thus:

$$\Delta S = nR \log_e \frac{V_2}{V_1} \tag{10.2a}$$

and

$$\Delta S = nR \log_e \frac{P_1}{P_2} \tag{10.3a}$$

For the purposes of discussing chemical equilibrium, it is often more useful to rewrite (10.3) in terms of S^0, the standard molar entropy. Consider, for instance, the generalized change:

$$1 \text{ mole gas } (1 \text{ atm}) \rightarrow 1 \text{ mole gas } (P \text{ atm})$$

For such a change, in view of equation (10.3),

$$\Delta S = S_2 - S_1 = R \log_e \frac{1}{P} = -R \log_e P$$

where the two pressures, 1 and P, are both in atmospheres. Now $S_1 = S^0$, the molar entropy at one atmosphere, so that

$$S_2 - S^0 = -R \log_e P$$

or

$$S = S^0 - R \log_e P \tag{10.4}$$

where S has been written in place of S_2, and P is in atmospheres.

Equation (10.4) gives us the entropy of a mole of gas at any pressure in terms of the standard molar entropy at a given temperature. Since the entropy is an extensive property, we can similarly write, for n moles, that

$$S = n \left\{ S^0 - R \log_e P \right\} \tag{10.5}$$

As an example of the use of equation (10.5), let us calculate the entropy of 0.683 moles of Xe at 25°C and 6.30 mm pressure. We must first express the pressure in atmospheres. We have

$$P = \frac{6.30}{760} = 8.29 \times 10^{-3} \text{ atm}$$

From Appendix 3 we find that for xenon $S^0 = 40.5$ cal deg^{-1}. Substituting into equation (10.5), we have:

$$S = n \left\{ S^0 - R \log_e P \right\}$$

$$= 0.683 \left\{ 40.5 - 1.987 \times 2.303 \log_{10} (8.29 \times 10^{-3}) \right\} \text{cal deg}^{-1}$$

where we have used the facts that $\log_e x = 2.303 \log_{10} x$ and that $R = 1.987$ cal deg^{-1}. Thus

$$S = 0.683 \left\{ 40.5 - 1.987 \times 2.303 (-2.0815) \right\} \text{cal deg}^{-1}$$

$$= 0.683 \left\{ 40.5 + 9.5 \right\} \text{cal deg}^{-1}$$

$$= 34.2 \text{ cal deg}^{-1}$$

We can also write equations equivalent to (10.4) and (10.5), for the enthalpy of an ideal gas. Since the enthalpy of an ideal gas does not vary with pressure or volume, we can write for one mole of gas at any temperature that:

$$H = H^0 \tag{10.6}$$

where H^0 is the standard molar enthalpy of the gas. For n moles of gas the equation is

$$H = nH^0 \tag{10.7}$$

These equations immediately make it possible to relate the free energy of a gas at constant temperature to the standard free energy, G^0. Combining (10.4) and (10.6) we have:

$$H - TS = (H^0 - TS^0) + RT \log_e P$$

or

$$G = G^0 + RT \log_e P \tag{10.8}$$

where G^0 is the standard molar free energy of the gas, i.e., the value at one atmosphere pressure. Similarly, for n moles we have:

$$G = n G^0 + n RT \log_e P \tag{10.9}$$

which gives us the free energy of a gas in terms of its standard molar free energy, its temperature, and its pressure.

THE ENTROPY AND ENTHALPY OF MIXTURES OF GASES

When several gases are allowed to mix without reacting chemically, the thermodynamic changes which occur are similar to those just discussed. If the intermolecular forces between like molecules as well as between unlike

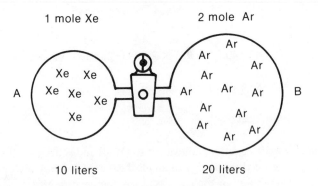

Figure 10-3. Container A holds one mole of xenon in a volume of 10 liters; container B holds two moles of argon in a volume of 20 liters. Though the tap is closed, the pressure is the same in both containers.

molecules are negligible, we find that each gas behaves thermodynamically as though the other gases were not present. Such a mixture of gases is said to be an *ideal gas mixture.*

Consider the situation depicted in Figure 10-3. One mole of xenon is in container A of volume 10 liters, while two moles of argon are in container B of volume 20 liters. The two gases are separated by a tap. Assuming the temperature to be the same in both containers, it follows that the pressures are also the same.

When the tap between the two containers is opened, xenon gas diffuses from A into B and argon gas diffuses in the reverse direction. An equilibrium state is soon obtained, consisting of a uniform mixture of both gases. One third of the mixture will be xenon and two thirds will be argon. Such a state of affairs is indicated diagrammatically in Figure 10-4. In accordance with our usual convention, we shall refer to the initial state shown in Figure 10-3 as State 1 and to the final state shown in Figure 10-4 as State 2.

We must first notice that there is no difference in internal energy between State 1 and State 2. On the assumption that there are no attractive or repulsive forces between like or unlike molecules*, it makes no difference to the potential energy of any of the xenon molecules whether they are surrounded by other xenon molecules as in State 1 or by a mixture of xenon and argon molecules as in State 2. The same is true of all the argon molecules. There is thus no change in the total potential energy of the molecules on mixing. Neither is there any change in the total kinetic energy. Collisions between xenon and argon molecules are as "elastic" as collisions between two xenon or two argon molecules. No kinetic energy is lost in any of the three kinds of collisions.

Since neither the total potential nor the total kinetic energy of the molecules changes, the internal energy does not change. Thus ΔE for the mixing of gases

---

*In the argument which follows we shall refer to xenon and argon *molecules* rather than to xenon and argon *atoms.* The two gases we have chosen as an example of mixing happen to have monatomic molecules. In most gaseous mixtures, though, the molecules will be diatomic or polyatomic. This use of the term molecule suggests (correctly) that the arguments used here apply more generally than just to the monatomic example actually under discussion.

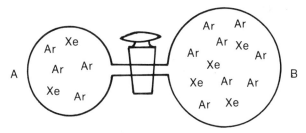

Figure 10-4. When the tap is opened, the gases mix so that the pressure, temperature, and composition are uniform in both containers.

must be zero. That ΔH is also zero is now easily argued. If the kinetic energy of the molecules is unaltered by mixing, so is the temperature. In consequence the pressure, is unaltered by mixing. (The total volume of the gases, of course, remains unchanged.) Since neither P nor V alters, their product PV remains the same as the system moves from State 1 to State 2; i.e., $\Delta(PV) = 0$.

Now

$$\Delta H = \Delta(E + PV)$$

$$= \Delta E + \Delta(PV)$$

but since both ΔE and $\Delta(PV)$ are zero,

$$\Delta H = 0$$

There is thus *no enthalpy change on the mixing of gases* provided only that the molecules do not attract or repel each other.

Although the enthalpy and the internal energy do not change when ideal gases mix, the same is not true of the entropy. As we have already seen in the last chapter, it increases. Let us now calculate the magnitude of this entropy change by deriving the relative probabilities of State 1 and State 2 by a method very similar to that just employed in the case of gaseous expansion.

The probability of a given xenon molecule being in container A is $1/3$, exactly as previously. By an identical argument the probability of finding a *mole* of xenon molecules *all* in container A rather than dispersed throughout both containers is again $(1/3)^N$. A similar argument applied to the 2N molecules of argon shows that the probability of finding all 2N of these in container B is $(2/3)^{2N}$.

The independence of the behavior of the two kinds of molecules allows us to use the law of *multiplication of independent probabilities*. Since the probability of finding the xenon molecules completely segregated in A rather than uniformly dispersed is $(1/3)^N$, and the probability of finding the argon molecules completely segregated in B rather than uniformly dispersed is $(2/3)^{2N}$, the product of these two probabilities, $(1/3)^N \times (2/3)^{2N}$, represents the probability of *both* argon *and* xenon being segregated in their separate containers rather than both being uniformly distributed throughout both containers. This product of two probabilities is the probability of State 1 relative to that of State 2. As pointed out previously, we can conclude that

$$\frac{W_1}{W_2} = \left(\frac{1}{3}\right)^N \left(\frac{2}{3}\right)^{2N}$$

Taking reciprocals, we then find that

$$\frac{W_2}{W_1} = 3^N \left(\frac{3}{2}\right)^{2N}$$

When natural logarithms are taken:

$$\log_e W_2 - \log_e W_1 = N \log_e 3 + 2N \log_e \frac{3}{2}$$

Multiplying by the Boltzmann constant, we obtain:

$$k \log_e W_2 - k \log_e W_1 = kN \log_e 3 + 2kN \log_e \frac{3}{2}$$

$$= R \log_e 3 + 2R \log_e \frac{3}{2}$$

so that

$$\Delta S = S_2 - S_1 = k \log_e W_2 - k \log_e W_1$$

$$= R \log_e 3 + 2R \log_e \frac{3}{2} \qquad \textbf{(10.10)}$$

$$= 3.79 \text{ cal deg}^{-1}$$

We should notice that the increase of entropy given by equation (10.10) is the same as we would have obtained had we considered the increase in entropy of each gas expanding independently of the other from its own container into the total volume of the two containers. The entropy change when a mole of xenon gas expands from 10 to 30 liters is given by equation (10.2). We find:

$$\Delta S_{Xe} = R \log_e \frac{V_2}{V_1} = R \log_e 3 \qquad \textbf{(10.11)}$$

Similarly, the entropy change when *two* moles of argon gas expand from 20 to 30 liters is given by equation (10.2a) as:

$$\Delta S_{Ar} = 2R \log_e \frac{V_2}{V_1} = 2R \log_e \frac{3}{2} \qquad \textbf{(10.12)}$$

On adding (10.11) and (10.12), we obtain (10.10):

$$\Delta S = \Delta S_{Xe} + \Delta S_{Ar} = R \log_e 3 + 2R \log_e \frac{3}{2}$$

The total entropy change can thus be regarded as the sum of two changes: (a) the entropy change when the xenon expands to fill the whole volume, and (b) the entropy change when the argon expands to fill the whole volume.

This is a particular example of a general rule which states, in a manner similar to Dalton's law of partial pressures, that the entropy of a mixture of gases is equal to the sum of the entropies each constituent would have if it occupied the entire volume at that temperature by itself. The explanation of this rule is quite simple. If we assume that the molecules of the constituent gases in the mixture behave *independently,* then their probabilities must be multiplied. The corresponding *entropies,* which are proportional to the *logarithms* of the probabilities, will then be *added.*

As an example of an entropy calculation involving this rule, let us calculate the total entropy of the N_2O_4-NO_2 equilibrium mixture at $25°C$ and 1 atmosphere pressure, which we discussed in the previous chapter. This mixture corresponds to the point Z in Figure 9-4 and consists of 0.814 moles of N_2O_4 and 0.372 moles of NO_2. The most direct method of performing this calculation is through the use of partial pressures.

The total number of moles of gas in this mixture is 0.814 + 0.372 = 1.186 moles. This number of moles of gas exerts a total pressure of one atmosphere. Since the partial pressure of a gas is equal to its mole fraction times the total pressure, the partial pressure of N_2O_4 must be 0.814/1.186 $\times$ 1 atm = 0.686 atm, while that of NO_2 is 0.372/1.186 $\times$ 1 atm = 0.314 atm.

We can now calculate the entropy of each gas separately, using the relationship given in equation (10.5):

$$S = n\left\{S^0 - R \log_e P\right\}$$

The value of the pressure to be used in each case is the pressure that each gas would have if it occupied the total volume by itself, or its partial pressure.

For N_2O_4, we find from Appendix 3 that $S^0 = 72.7$ cal deg^{-1} mole^{-1}. Thus

$$S_{N_2O_4} = 0.814 \left\{72.7 \text{ cal deg}^{-1} - 1.987 \text{ cal deg}^{-1} \times 2.303 \log_{10} 0.686\right\}$$

$$= 0.814 \left\{72.7 + 0.7\right\} \text{ cal deg}^{-1}$$

$$= 59.7 \text{ cal deg}^{-1}$$

For NO_2, we find that $S^0 = 57.3$ cal deg^{-1} mole^{-1}, and we have:

$$S_{NO_2} = 0.372 \left\{57.3 \text{ cal deg}^{-1} - 1.987 \text{ cal deg}^{-1} \times 2.303 \log_{10} 0.314\right\}$$

$$= 0.372 \left\{57.3 + 2.3\right\} \text{ cal deg}^{-1}$$

$$= 22.2 \text{ cal deg}^{-1}$$

Adding these two results then gives the entropy of the mixture:

$$S = 81.9 \text{ cal deg}^{-1}$$

In general, if we mix a moles of gas A, b moles of gas B, c moles of gas C, etc., the entropy of the final mixture is given by a sum of entropy terms, one for each gas. Each of these terms corresponds to the entropy which that particular gas would have if it occupied the final volume on its own. Thus, if S^0 (A) is the standard molar entropy of constituent A, and P_A is its partial pressure, the contribution of A to the total entropy is given by

$$S(A) = a\left\{ S^0(A) - R \log_e P_A \right\} \tag{10.13a}$$

and in a similar way

$$S(B) = b\left\{ S^0(B) - R \log_e P_B \right\} \tag{10.13b}$$

$$S(C) = c\left\{ S^0(C) - R \log_e P_C \right\} \tag{10.13c}$$

and so on.

The total entropy is then given by the sum of these terms; i.e.,

$$S = S(A) + S(B) + S(C) + \ldots\ldots\ldots$$

or

$$S = a\left\{ S^0(A) - R \log_e P_A \right\} + b\left\{ S^0(B) - R\log_e P_B \right\} \\ + c\left\{ S^0(C) - R \log_e P_C \right\} + \ldots\ldots\ldots\ldots \tag{10.14}$$

The total *enthalpy* of a gaseous mixture can similarly be regarded as the sum of the individual enthalpies each of the constituent gases would have if it occupied the given volume by itself. The resultant expression for the total enthalpy is, however, much simpler than equation (10.14) due to the fact that enthalpies of ideal gases and ideal gas mixtures show no dependence on pressure. Accordingly, for a mixture of a moles of A, b moles of B, c moles of C, etc., the enthalpy is given by:

$$H = aH^0(A) + bH^0(B) + cH^0(C) + \ldots\ldots \tag{10.15}$$

where $H^0(A)$, $H^0(B)$, $H^0(C)$, etc., are the standard molar enthalpies of the gases involved.

THE FREE ENERGY OF MIXTURES OF GASES

Equations (10.14) and (10.15) can be combined to yield a general equation for the free energy of a mixture of gases. Suppose, as before, that we have a mixture of a moles of gas A, b moles of gas B, c moles of gas C, etc. The free energy, G, of such a mixture is given by the equation:

$$G = H - TS \qquad \textbf{(10.16)}$$

where H is given by equation (10.15) and S by equation (10.14). Substituting these into (10.16) we then have:

$$G = H - TS = a \left\{ H^0(A) - TS^0(A) + RT \log_e P_A \right\}$$
$$+ b \left\{ H^0(B) - TS^0(B) + RT \log_e P_B \right\} + c \left\{ H^0(C) - TS^0(C) + RT \log_e P_C \right\} + \ldots$$

or, since $G^0 = H^0 - TS^0$,

$$G = a \left\{ G^0(A) + RT \log_e P_A \right\} + b \left\{ G^0(B) + RT \log_e P_B \right\}$$
$$+ c \left\{ G^0(C) + RT \log_e P_C \right\} + \ldots \ldots \qquad \textbf{(10.17)}$$

We can write (10.17) as

$$G = G(A) + G(B) + G(C) + \ldots \ldots \qquad \textbf{(10.18)}$$

where

$$G(A) = a \left\{ G^0(A) + RT \log_e P_A \right\}$$

$$G(B) = b \left\{ G^0(B) + RT \log_e P_B \right\}$$

and so on. In other words, the total free energy of the mixture, G, is equal to the sum of the individual free energies, $G(A)$, $G(B)$, etc. that the gases would have if each occupied the given volume on its own. Manipulating the free energy of a gaseous mixture is thus not much more difficult than manipulating the pressure of such a mixture by adding the partial pressures in accordance with Dalton's Law.

As an example of the use of equation (10.17), let us derive a formula to calculate the free energy changes shown in Figure 9-4; i.e., to calculate how the free energy varies with the extent of a reaction during the conversion of one mole of N_2O_4 to two moles of NO_2 according to the reaction

$$N_2O_4(g) \rightleftharpoons 2NO_2(g) \quad (25°C, 1 \text{ atm})$$

As we discovered in Chapter 9 (p. 192), ΔG^0 for this reaction has the value +1.15 kcal.

Suppose that x moles of N_2O_4 have decomposed in this reaction, leaving $(1 - x)$ moles of this species and producing 2x moles of NO_2. The total number of moles is then $(1-x) + 2x = (1+x)$ moles. Since the total pressure is 1 atmosphere throughout the reaction, we can now calculate the partial pressure of each constituent. That of N_2O_4 is

$$P_{N_2O_4} = \text{mole fraction} \times \text{total pressure}$$

$$= \frac{(1-x)}{(1+x)} \, 1 \text{ atm}$$

while that of NO_2 is

$$P_{NO_2} = \frac{2x}{1+x} \text{ atm}$$

Substituting these two partial pressures into equation (10.17), we then obtain:

$$G = (1-x)\left\{G^0(N_2O_4) + RT \log_e P_{N_2O_4}\right\} + 2x\left\{G^0(NO_2) + RT \log_e P_{NO_2}\right\}$$

$$= (1-x)\left\{G^0(N_2O_4) + RT \log_e\left(\frac{1-x}{1+x}\right)\right\} + 2x\left\{G^0(NO_2) + RT \log_e\left(\frac{2x}{1+x}\right)\right\}$$

$$= G^0(N_2O_4) + x\left\{2G^0(NO_2) - G^0(N_2O_4)\right\}$$
$$+ RT\left\{(1-x)\log_e\left(\frac{1-x}{1+x}\right) + 2x\log_e\left(\frac{2x}{1+x}\right)\right\}$$

$$= G^0(N_2O_4) + x\Delta G^0 + RT\left\{(1-x)\log_e\left(\frac{1-x}{1+x}\right) + 2x\log_e\left(\frac{2x}{1+x}\right)\right\} \qquad \textbf{(10.19)}$$

For simplicity we shall make the standard molar free energy of N_2O_4 our arbitrary zero of energy. We can then substitute

$$G^0(N_2O_4) = 0.0$$

into (10.19) and obtain

$$G = 1.15x \text{ kcal} + RT\left\{(1-x)\log_e\left(\frac{1-x}{1+x}\right) + 2x \log_e\left(\frac{2x}{1+x}\right)\right\} \qquad \textbf{(10.20)}$$

Equation (10.20) enables us to calculate the value of G for any value of x. For instance, if $x = 0.20$, then

$$G = 0.2 \times 1.15 \text{ kcal} + RT \left\{ 0.8 \log_e \left(\frac{0.8}{1.2} \right) + 0.4 \log_e \left(\frac{0.4}{1.2} \right) \right\}$$

$$= 0.230 \text{ kcal} + 1.987 \times 10^{-3} \text{ kcal deg}^{-1} \times 298.2 \text{ deg}$$

$$\times 2.303 \left\{ 0.8 \log_{10} \left(\frac{2}{3} \right) + 0.4 \log_{10} \left(\frac{1}{3} \right) \right\}$$

$$= 0.230 \text{ kcal} - 0.453 \text{ kcal}$$

$$= -0.223 \text{ kcal}$$

$$= -223 \text{ cal}$$

Proceeding in this fashion for other values of x enables us to draw the curve AZB, describing the variation of G with the extent of reaction, in Figure 9-4. By drawing such a curve we can find the value of x corresponding to the minimum value of G, and hence the composition of the equilibrium mixture of NO_2 and N_2O_4.

This method for finding the minimum value for G is a little laborious. There are better ways. Those who are familiar with the calculus will realize that the position of the minimum can be obtained by differentiating equation (10.20) with respect to x and equating the result to zero. Why not perform this operation and check that the minimum corresponds to the point x = 0.186?

There is an even easier method of finding the equilibrium composition; this is by using the *equilibrium constant* K_p. The connection between K_p and the free energy G will constitute the subject matter of our final chapter.

PROBLEMS

1. Two containers, A and B, are separated by a tap as in Figure 10-1. A has a volume of 15 liters and contains one mole of ideal gas at 25°C. B has a volume of 45 liters and is completely evacuated. Calculate the change in entropy which occurs when the tap is opened. What is the corresponding change in free energy?

2. In the previous problem, the system loses free energy when the tap is opened. What happens to the free energy which is lost?

3. If an ideal gas is suddenly compressed, does its temperature necessarily increase?

4. Three molecules of neon are added to the system shown in Figure 10-4. What is the probability of finding one of these in container A and two in container B? What is the probability of finding all three in container A?

5. Two containers, A and B, are separated by a tap as in Figure 10-1. A has a volume of 15 liters and contains one mole of N_2. B has a volume of 30 liters and contains two moles of N_2. Calculate the change in entropy which occurs when the tap is opened.

6. In the previous problem, if A and B each contained one mole of N_2, what would the change in entropy be?

7. Calculate the entropy of a mixture of two moles of CO gas and three moles of CO_2 gas at $25°C$ and 1 atmosphere pressure.

8. Calculate ΔS, ΔH, and ΔG for the following process at $25°C$ and at $500°K$:

1 mole H_2(1 atm) + 2 mole CH_4(1 atm) → 3 mole mixture(1 atm)

9. Fill in the blanks in the table given below by using equation (10.20) to calculate the free energy of the reaction

$$N_2O_4(g) \rightarrow 2NO_2(g) \quad (25°C, 1 \text{ atm})$$

for various values of x, the extent of the reaction. Then draw a graph showing the variation of G with x.

Extent of Reaction x	Free Energy G(calories)
0.000	0.00
0.100	
0.186	
0.200	−223
0.300	
0.400	−107
0.500	
0.600	+157
0.700	
0.800	+548
0.900	
1.000	+1150

10. Show that the expression

$$G = G^0 + RT \left\{ x \log_e x + (1-x)\log_e(1-x) \right\}$$

describes the variation of free energy with extent of reaction when one mole of an optically active gas, such as the isomer A shown in Figure 4-1, is converted into one mole of its other optical isomer. G^0 is the standard molar free energy of either isomer and x is the extent of the reaction.

11. If you are able to differentiate, show that the minimum free energy in the previous problem corresponds to x = 0.5.

12. The remark is sometimes made that all entropy is entropy of mixing. The following problem gives some idea of the significance of this remark. In Problem 19 of Chapter 6 we calculated the entropy of a mole of particles, one third of which were in the first excited state and the other two thirds of which were in the ground state. Show that the same result can be obtained by calculating the entropy of mixing of one third of a mole of excited molecules and two thirds of a mole of unexcited molecules.

Chapter 11

FREE ENERGY AND EQUILIBRIUM

THE EQUILIBRIUM BETWEEN A LIQUID AND ITS VAPOR

The standard molar free energy of formation of liquid water at $25°C$ is -56.687 kcal, while that for the vapor at the same temperature is slightly more positive, namely -54.634 kcal. These two ΔG_f^0 values not only enable us to calculate that the value of ΔG^0 for the reaction

$$H_2O(g) \rightarrow H_2O(l) \quad (25°C, 1 \text{ atm})$$

is -2.053 kcal; they also tell us that the reaction will go to completion, provided the pressure is kept constant. Every least bit of H_2O vapor at one atmosphere pressure which converts to liquid water lowers the free energy. This process of lowering the free energy can only stop when all the vapor has been converted to liquid. In Figure 11-1 we show how the free energy varies as one mole of gaseous H_2O is converted into one mole of liquid H_2O, while temperature and pressure remain constant at $25°C$ and 1 atmosphere.

Figure 11-1 also illustrates how the free energy changes during the same reaction at the same temperature of $25°C$, but at a much lower pressure. The pressure in question is 23.8 mm of mercury. This is the vapor pressure of water at $25°C$. At this pressure and temperature the molar free energy of pure liquid and pure vapor are the same, so that as one mole of vapor changes into one mole of liquid, no free energy change occurs. This means that we can mix any quantity of liquid water with any quantity of water vapor at this pressure and temperature and have an equilibrium state in which there is no net tendency for vapor to turn into liquid or vice versa. The condition for equilibrium between water and its vapor is thus quite a simple one: both phases must have the *same molar free energy.*

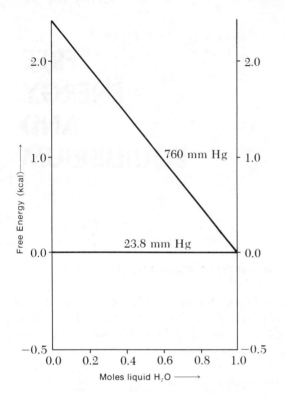

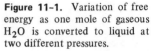

Figure 11-1. Variation of free energy as one mole of gaseous H_2O is converted to liquid at two different pressures.

The argument just employed obviously applies in the general case of liquid-vapor equilibrium. If a liquid and its vapor are in equilibrium, and $G(g)$ is the molar free energy of the vapor while $G(l)$ is the molar free energy of the liquid, then we can write:

$$G(l) = G(g) \qquad \qquad (11.1)$$

If P is the vapor pressure of the vapor we have the relationship given by equation (10.9):

$$G(g) = G^0(g) + RT \log_e P$$

where G^0 is the standard molar free energy of the vapor, the value of $G(g)$ at a pressure of one atmosphere. Furthermore, since the free energy of a liquid (or a solid) changes very slowly with pressure, we can also write:

$$G(l) = G^0(l) \qquad \qquad (11.2)$$

Thus we can assume that the free energy of the liquid has the same value at pressure P as it has at one atmosphere pressure.

Substituting (11.2) and (10.9) into (11.1), we have the result

$$G^0(l) = G^0(g) + RT \log_e P$$

so that

$$G^0(g) - G^0(l) = -RT \log_e P$$

and finally

$$\Delta G^0 = -RT \log_e P \qquad \text{(11.3)}$$

where ΔG^0 is the standard free energy change for the vaporization process:

$$1 \text{ mole liquid} \rightarrow 1 \text{ mole vapor } (T^\circ C, 1 \text{ atm})$$

Equation (11.3) allows us to calculate the vapor pressure of a liquid from free energy data. As an example of this calculation, let us apply the equation to water vapor at $25^\circ C$. For the process

$$H_2O(l) \rightarrow H_2O(g) \quad (25^\circ C, 1 \text{ atm})$$

ΔG^0, as calculated at the beginning of the chapter, has the value +2.053 kcal. Substituting this value into (11.3) we have

$$\Delta G^0 = 2.053 \text{ kcal} = 2053 \text{ cal}$$

$$= -RT \log_e P$$

$$= -1.987 \text{ cal deg}^{-1} \times 298.2 \text{ deg} \times 2.303 \log_{10} P$$

$$= -1370 \text{ cal} \times \log_{10} P$$

Thus

$$\frac{2053 \text{ cal}}{-1370 \text{ cal}} = -1.500 = \log_{10} P$$

Hence

$$P = \text{antilog } (0.500 - 2)$$

$$= 3.16 \times 10^{-2} \text{ atm}$$

$$= 3.16 \times 10^{-2} \text{ atm} \times \frac{760 \text{mm Hg}}{1 \text{ atm}}$$

$$= 24.0 \text{ mm Hg}$$

The measured value is 23.756 mm. The small discrepancy is due to the non-ideal behavior of H_2O gas.

ISOMERIZATION EQUILIBRIA

The next equilibrium to be considered is also one in which there is a single reactant and a single product. We will again find, although the circumstances are different, that the equilibrium position corresponds to the point where the molar free energy of the reactant is equal to that of the product.

The equilibrium we will discuss is that between the two gaseous isomers of butane, C_4H_{10}:

$$\begin{matrix} H_3C \\ \diagdown \\ H_3C \diagup \end{matrix} CH-CH_3 \quad \rightleftharpoons \quad CH_3-CH_2-CH_2-CH_3 \qquad \text{(11.4)}$$

isobutane   n-butane

at $25°C$ and one atmosphere pressure. Both isomers have the same molecular formula, but isobutane has a branched chain of carbon atoms, while n-butane has a "straight" chain.

The value of ΔG_f^0 for isobutane is -4.296 kcal, while that for n-butane is -3.754 kcal. The standard free energy change, ΔG^0, for reaction (11.4) thus has the rather small value of 0.542 kcal. Accordingly, we expect that there will be comparable quantities of the two isomers present at equilibrium at this temperature and pressure. This prediction is borne out in practice. In Figure 11-2 the variation of free energy as one mole of isobutane is converted into one mole of n-butane is illustrated. A minimum occurs at X, where 0.286 mole of n-butane and 0.714 mole of isobutane are in equilibrium with each other.

How can we predict the value of this minimum? In order to answer this question we must first investigate how the molar free energy of each species behaves as the reaction proceeds. Consider first the molar free energy of

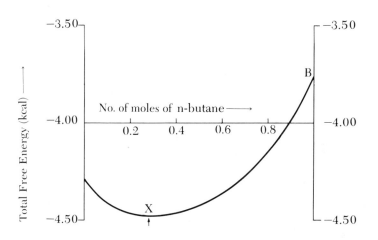

Figure 11-2. The variation in free energy as one mole of isobutane is converted into n-butane at one atmosphere and 25°C.

isobutane. This is given by

$$G(\text{isobutane}) = G^0(\text{isobutane}) + RT \log_e P_{ib} \qquad \textbf{(11.5a)}$$

where P_{ib} is the partial pressure of isobutane. A similar equation holds for the other isomer:

$$G(\text{n-butane}) = G^0(\text{n-butane}) + RT \log_e P_{nb} \qquad \textbf{(11.5b)}$$

If we let x be the number of moles of n-butane formed at any given stage of the reaction, then the reaction mixture will comprise x moles of n-butane and (1 − x) moles of isobutane, a total of 1 mole of molecules.

Now, the partial pressure of a gas is equal to its mole fraction times the total pressure. We can thus write for the partial pressure of isobutane:

$$P_{ib} = \left(\frac{1-x}{1}\right)(1 \text{ atm}) = (1-x) \text{ atm} \qquad \textbf{(11.6a)}$$

while the partial pressure of the straight chain isomer is given by

$$P_{nb} = \left(\frac{x}{1}\right)(1 \text{ atm}) = x \text{ atm} \qquad \textbf{(11.6b)}$$

Substituting these results into equations (11.5a) and (11.5b), we then obtain

$$G(\text{isobutane}) = G^0(\text{isobutane}) + RT \log_e(1-x) \qquad \textbf{(11.7a)}$$

and

$$G(\text{n-butane}) = G^0(\text{n-butane}) + RT \log_e(x) \qquad \textbf{(11.7b)}$$

In Figure 11-3 the way in which these two molar free energies behave as the reaction proceeds is shown. G(isobutane) decreases as the reaction proceeds *because its mole fraction and hence its partial pressure decreases.* G(n-butane), on the other hand, *increases as its partial pressure builds up.*

During the early stages of the reaction, before the point Y is reached, the molar free energy of isobutane is higher than that of n-butane. Under these conditions, if a small amount of isobutane, say a nanomole*, is converted into n-butane, the free energy must *decrease,* because the free energy of a nanomole of n-butane is less than the free energy of the nanomole of isobutane which it replaces. As can be seen in Figure 11-2, the total free energy of the reaction mixture does decrease to start with. Returning to Figure 11-3 we notice that

*1 nanomole = 10^{-9} mole

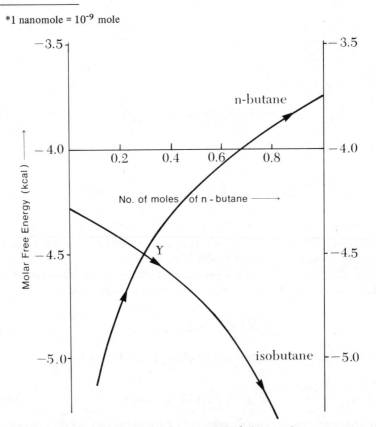

Figure 11-3. Behavior of the molar free energies of the two isomers as one mole of isobutane is converted into one mole of n-butane at 25°C and one atmosphere total pressure.

after the point Y is passed we have the reverse situation to that just considered. Here it is the *product* which has the higher molar free energy. As the reaction proceeds beyond Y, each nanomole of isobutane (which now has a lower free energy) is replaced by a nanomole of n-butane of higher free energy. Beyond point Y, therefore, the free energy must increase as the reaction proceeds. This behavior is shown in Figure 11-2.

It is now easy to see what happens at the point Y itself, when the molar free energies of reactant and product are equal. This is the exact point at which the total free energy stops decreasing and starts increasing. It corresponds to the minimum in the free energy curve, point X, of Figure 11-2. Both this point and point Y in Figure 11-3 thus correspond to the same equilibrium composition.

The condition for equilibrium in this instance is thus the same as in our first example. The molar free energy of *reactant* and *product* must have the *same value* at equilibrium. In mathematical terms,

$$G(\text{isobutane}) = G(\text{n-butane}) \tag{11.8}$$

We can use equation (11.8) to determine the value of x, the number of moles of n-butane present at equilibrium. Substituting equations (11.5a) and (11.5b) into (11.8), we have

$$G^0(\text{isobutane}) + RT \log_e P_{ib} = G^0(\text{n-butane}) + RT \log_e P_{nb}$$

or

$$G^0(\text{n-butane}) - G^0(\text{isobutane}) = RT \left\{ \log_e P_{ib} - \log_e P_{nb} \right\}$$

i.e.,

$$\Delta G^0 = - RT \log_e \frac{P_{nb}}{P_{ib}} \tag{11.9}$$

Substituting in the values for the partial pressures (equations (11.6a) and (11.6b)), we then have

$$\Delta G^0 = - RT \log_e \left(\frac{x}{1-x} \right)$$

Since $\Delta G^0 = +542$ calories, we have

$$542 \text{ cal} = - 1.987 \text{ cal deg}^{-1} \times 298.2 \text{ deg} \times 2.303 \log_{10} \left(\frac{x}{1-x} \right)$$

i.e.,

$$\log_{10}\left(\frac{x}{1-x}\right) = \frac{-542}{1.987 \times 298.2 \times 2.303} = -0.397$$

$$= 0.603 - 1$$

Thus

$$\left(\frac{x}{1-x}\right) = \text{antilog}\,(0.603 - 1) = 0.401$$

or
$$x = 0.401 - 0.401x.$$

so that, finally,

$$x = 0.286$$

There are thus 0.286 mole of n-butane and 0.714 mole of isobutane in the equilibrium mixture.

We can now generalize our above arguments to apply to any reaction of the form

$$A(g) \rightleftharpoons B(g) \tag{11.10}$$

The condition for equilibrium for any reaction of this type will be that the molar free energies of A and B have the same value, i.e.,

$$G(A) = G(B)$$

or

$$G^0(A) + RT\,\log_e P_A = G^0(B) + RT\,\log_e P_B \tag{11.11}$$

where P_A and P_B are the partial pressures of the two species involved. Rearranging equation (11.11), we then have

$$G^0(B) - G^0(A) = RT\,\log_e P_A - RT\,\log_e P_B$$

or

$$\Delta G^0 = -RT\,\log_e \frac{P_B}{P_A} \tag{11.12}$$

an equation which will enable us to calculate the equilibrium position for any isomerization reaction of the type given by equation (11.10).

An important thing to notice about equation (11.12) is that it relates a free energy change to an *equilibrium constant*. The pressure equilibrium constant*, K_p, for reaction (11.10) is given by the expression:

$$K_p = \frac{P_B}{P_A} \qquad \textbf{(11.13)}$$

Substituting this relationship into equation (11.12) yields the result

$$\Delta G^0 = -RT \log_e K_p \qquad \textbf{(11.14)}$$

Equation (11.14) relates the free energy difference between reactants and products (at standard one atmosphere pressure) to the equilibrium constant. It is one of the more important relationships in chemical thermodynamics. As we shall see very shortly, it applies not only to isomerization reactions, but to *all* gaseous reactions.

THE $N_2O_4 \rightleftharpoons 2NO_2$ EQUILIBRIUM AGAIN

The two equilibria we have so far considered in this chapter have both been of the rather simple form

$$A \rightleftharpoons B$$

In both cases we found that the condition for chemical equilibrium was that the molar free energy of the reactant be equal to the molar free energy of the product. In more complicated equilibria, such a simple criterion is no longer adequate. For example, in the reaction

$$N_2O_4(g) \rightleftharpoons 2NO_2(g) \quad (25°C, 1 \text{ atm}) \qquad \textbf{(11.15)}$$

the equilibrium state does *not* correspond to a mixture in which the molar free energy of N_2O_4 is equal to the molar free energy of NO_2.

To see why this is so, let us again consider Figure 9-4. In this figure the way in which the free energy varies as one mole of N_2O_4 decomposes into two moles of NO_2 is plotted. The free energy at first decreases, attains a minimum, and then increases. The free energy changes in this way as the reaction proceeds for much the same reason as that noted in the isomerization reaction. Initially, the loss in free energy which occurs as the reactant decomposes is larger than the

---

*Readers unfamiliar with the quantity K_p should consult Appendix 2.

gain in free energy due to the formation of product. As the reaction proceeds, though, the reverse becomes true.

There is one important difference between this case and the previous case. Whereas in the isomerization reaction each mole of reactant was converted into *one* mole of product, in this reaction one mole of reactant is converted into *two* moles of product. It is this fact which leads to a different criterion for equilibrium in the two cases.

Let us imagine the decomposition of one mole of N_2O_4, (Figure 9-4) as occurring in small stages, a nanomole at a time. In the early part of the reaction the free energy of each nanomole of N_2O_4 which decomposes is *larger* than that of the two nanomoles of NO_2 which are formed. As a result, the total free energy decreases as the N_2O_4 decomposes. Eventually, though, the total free energy starts to increase. This increase only occurs when the free energy of each nanomole of N_2O_4 which is destroyed is *less* than that of the two nanomoles of NO_2 which are produced.

As the free energy stops decreasing and starts to increase, it passes through a minimum. At the minimum the free energy of one nanomole of N_2O_4 in the reaction mixture must be *equal* to the free energy of two nanomoles of NO_2. In other words, the equilibrium state corresponds to a mixture in which the molar free energy of the N_2O_4 is equal to *twice* the molar free energy of the NO_2.

In mathematical terms we can write this condition as

$$G(N_2O_4) = 2G(NO_2) \tag{11.16}$$

Introducing standard values and partial pressures, this equation becomes

$$G^0(N_2O_4) + RT \log_e P_{N_2O_4} = 2 \left\{ G^0(NO_2) + RT \log_e P_{NO_2} \right\}$$

Rearranging, we then obtain

$$2G^0(NO_2) - G^0(N_2O_4) = -RT \log_e(P_{NO_2})^2 + RT \log_e P_{N_2O_4}$$

or

$$\Delta G^0 = -RT \log_e \frac{(P_{NO_2})^2}{P_{N_2O_4}} \tag{11.17}$$

Now, K_p for reaction (11.15) is given by

$$K_p = \frac{(P_{NO_2})^2}{P_{N_2O_4}} \tag{11.18}$$

Substituting this expression into (11.17) we again obtain the result of equation (11.14):

$$\Delta G^0 = -RT \log_e K_p \qquad (11.14)$$

We shall shortly prove that equation (11.14) is valid in the general case. Before we move on to such a proof, though, let us calculate the exact position of equilibrium when one mole of N_2O_4 decomposes into NO_2 at $25°C$ at a constant total pressure of one atmosphere. K_p is first calculated from ΔG^0, by substituting the value of ΔG^0_{298} for the reaction into equation (11.14). We then have:

$$\Delta G^0 = -RT \log_e K_p$$

$$1150 \text{ cal} = -1.987 \text{ cal deg}^{-1} \times 298.2 \text{ deg} \times 2.303 \log_{10} K_p$$

Thus

$$\log_{10} K_p = \frac{-1150}{1.987 \times 298.2 \times 2.303}$$

$$= -0.843 = 0.157 - 1$$

so that

$$K_p = \text{antilog} (0.157 - 1) = 1.435 \times 10^{-1}$$

$$= 0.1435$$

Substituting this result into (11.18) gives

$$\frac{(P_{NO_2})^2}{P_{N_2O_4}} = 0.1435 \qquad (11.19)$$

Suppose next that x is the number of moles of N_2O_4 that must decompose to achieve equilibrium. There will then be $(1-x)$ moles of N_2O_4 left in the equilibrium mixture. The decomposition of x moles of N_2O_4 results in the production of 2x moles of NO_2. The equilibrium mixture thus consists of a total of $(1+x)$ moles of molecules, $(1-x)$ mole being N_2O_4 and 2x mole being NO_2. Since the total pressure is 1 atmosphere, we can now write down expressions for the partial pressures of the two constituents of the equilibrium mixture:

$$P_{N_2O_4} = \left(\frac{1-x}{1+x}\right)(1 \text{ atm})$$

and

$$P_{NO_2} = \left(\frac{2x}{1+x}\right)(1 \text{ atm})$$

Substituting these two expressions into (11.19), we then obtain

$$0.1435 = \frac{\left(\dfrac{2x}{1+x}\right)^2}{\left(\dfrac{1-x}{1+x}\right)} = \frac{4x^2}{(1+x)(1-x)}$$

so that

$$\frac{4x^2}{1-x^2} = 0.1435$$

and

$$4.1435x^2 = 0.1435$$

$$x^2 = \frac{0.1435}{4.1435} = 0.0347$$

and finally

$$x = 0.186$$

The equilibrium mixture thus consists of $1-x = 0.814$ mole of N_2O_4 and $2x = 0.372$ mole of NO_2. These are the figures quoted in Chapter 9 and given in Figure 9-4.

GENERAL TREATMENT OF GASEOUS EQUILIBRIA

We can now consider the general case of a chemical equilibrium involving only gases. Suppose that a moles of compound A, b moles of compound B and so on, react to produce m moles of compound M, n moles of compound N, etc., according to the equation:

$$aA + bB + cC + \ldots \rightleftharpoons mM + nN + \ldots \qquad \textbf{(11.20)}$$

Let us further suppose that the reaction has proceeded until equilibrium has been reached, so that all the constituent gases, both reactants and products, are present at their equilibrium partial pressures. We can then write for the molar free energy of A that

$$G(A) = G^0(A) + RT \log_e P_A \qquad (11.21a)$$

where P_A is the partial pressure of A, and $G^0(A)$ is its standard molar free energy (i.e., the free energy of one mole of A at one atmosphere pressure). Similar expressions can be written for all the other constituents of the equilibrium mixtures:

$$\left.\begin{aligned}
G(B) &= G^0(B) + RT \log_e P_B \\[6pt]
G(C) &= G^0(C) + RT \log_e P_C \\[6pt]
G(M) &= G^0(M) + RT \log_e P_M \\[6pt]
G(N) &= G^0(N) + RT \log_e P_N
\end{aligned}\right\} \qquad (11.21b)$$

and so on.

Consider next what would happen to this equilibrium mixture in terms of free energy if it were altered very slightly as the forward reaction occurred to a very small extent. The behavior would be the same as for the two particular equilibria considered above. The loss in free energy accompanying the destruction of reagents would be exactly balanced out by the gain in free energy caused by the formation of products. Suppose that in such a small shift from the equilibrium position, a nanomoles of compound A, b nanomoles of compound B, etc. react to form m nanomoles of compound M, n nanomoles of compound N, etc. The loss in free energy due to destruction of reagents A, B, etc. is then

$$\Delta G_I = -10^{-9} \left\{ aG^0(A) + aRT \log_e P_A + bG^0(B) + bRT \log_e P_B + \ldots \right\}$$

while the gain in free energy due to the formation of products M, N, etc. is given by:

$$\Delta G_{II} = -10^{-9} \left\{ mG^0(M) + mRT \log_e P_M + nG^0(N) + nRT \log_e P_N + \ldots \right\}$$

Since we are at a free energy minimum, the loss in free energy ΔG_I must exactly balance out the gain in free energy ΔG_{II}:

$$\Delta G_I = \Delta G_{II}$$

so that

$$aG^0(A) + bG^0(B) + \ldots + RT \left\{ \log_e(P_A)^a + \log_e(P_B)^b + \ldots \right\}$$
$$= mG^0(M) + nG^0(N) + \ldots + RT \left\{ \log_e(P_M)^m + \log_e(P_N)^n + \ldots \right\}$$

Rearranging, we then have:

$$\{mG^0(M) + nG^0(N) + \ldots\} - \{aG^0(A) + bG^0(B) + \ldots\}$$
$$= -RT \log_e \{(P_M)^m \times (P_N)^n \times \ldots\} + RT \log_e \{(P_A)^a \times (P_B)^b \times \ldots\}$$

The right hand side of this equation is equal to the difference between the standard molar free energy of the products and that of the reactants; i.e., it is equal to ΔG^0. We can therefore write:

$$\Delta G^0 = RT \log_e \left\{ \frac{(P_M)^m \times (P_N)^n \times \ldots}{(P_A)^a \times (P_B)^b \times \ldots} \right\}$$

But the expression in brackets above is simply the general expression for the pressure equilibrium constant K_p, so we once again obtain equation (11.14),

$$\Delta G^0 = -RT \log_e K_p \qquad\qquad\qquad \textbf{(11.14)}$$

By proving equation (11.14) for the general case, we have not only established that the equilibrium constant of a reaction is related to a free energy change; we have also shown that the equilibrium constant really is a *constant* for a given temperature. We can rewrite equation (11.14) in the form

$$\frac{-\Delta G^0}{RT} = \log_e K_p \qquad\qquad\qquad \textbf{(11.22)}$$

It is easy to establish that the left-hand side of this equation depends only on temperature. The quantity R is a constant of nature, while ΔG^0 and T vary only with temperature. Hence $\log_e K_p$ and therefore K_p itself depend only on the temperature and are constant for a given temperature. This can only mean that in an equilibrium state, corresponding to a minimum free energy, the partial pressures of products and reactants *must* conform to the relationship:

$$\frac{(P_M)^m \times (P_N)^n \times \ldots}{(P_A)^a \times (P_B)^b \times \ldots} = K_p = \text{constant}$$

If the partial pressures of gases in a mixture under a particular set of conditions do not satisfy this relationship, then the free energy is not at a minimum and an equilibrium state has not yet been attained.

EQUILIBRIUM CALCULATIONS

Equation (11.14) allows us to predict an equilibrium constant for a reaction, and hence the equilibrium composition, from free energy data. How this is done is best explained by an example. Let us calculate how much NH_3 gas is produced when one mole of N_2 and 3 moles of H_2 are allowed to react at $400°C$ and a constant total pressure of ten atmospheres. In order to perform the calculation we need a value for ΔG^0 for the reaction

$$N_2(g) + 3H_2(g) \rightleftharpoons 2NH_3(g) \qquad\qquad (11.23)$$

at $400°C$. The best available value for this quantity is $+11.55$ kcal. We shall see a little later how this value is obtained. For the moment, though, let us take it as given and use it to calculate the equilibrium composition. Substituting $+11.55$ kcal into equation (11.14), we have:

$$\Delta G^0 = -RT \log_e K_p$$

$$11,550 \text{ cal} = -1.987 \text{ cal deg}^{-1} \times 673.2 \text{ deg} \times 2.303 \log_{10} K_p$$

or

$$\log_{10} K_p = -3.749 = 0.251 - 4$$

Thus

$$K_p = 1.782 \times 10^{-4}$$

In other words,

$$\frac{(P_{NH_3})^2}{(P_{N_2})(P_{H_2})^3} = 1.782 \times 10^{-4} \text{ at } 400°C \qquad\qquad (11.24)$$

Suppose next that, when one mole of N_2 and 3 moles of H_2 are equilibrated at this temperature and 10 atm pressure, x moles of N_2 react with 3x moles of H_2 to form 2x moles of NH_3. This leaves (1-x) moles of N_2 and (3-3x) moles of H_2 unreacted. The total number of moles of each species is thus:

$$N_2 \quad 1\text{-}x \ \text{moles}$$

$$H_2 \quad 3\text{-}3x \ \text{moles}$$

$$NH_3 \quad 2x \ \text{moles}$$

giving a total of 4- 2x moles.

We can now write down the partial pressure of each constituent in the mixture, using Dalton's law and remembering that the total pressure is 10 atmospheres:

$$P_{N_2} = \frac{(1-x)}{(4-2x)}(10 \text{ atm})$$

$$P_{H_2} = \frac{(3-3x)}{(4-2x)}(10 \text{ atm})$$

$$P_{NH_3} = \frac{2x}{(4-2x)}(10 \text{ atm})$$

Substituting these expressions into equation (11.24) we then obtain

$$1.782 \times 10^{-4} = \frac{\dfrac{(2x)^2 (10)^2}{(4-2x)^2}}{\dfrac{(1-x)(10)}{(4-2x)}\dfrac{(3-3x)^3 (10)^3}{(4-2x)^3}}$$

Simplifying, we find:

$$1.782 \times 10^{-4} = \frac{4x^2 (4-2x)^2}{10^2 (1-x)(3-3x)^3} = \frac{16x^2 (2-x)^2}{27 \times 10^2 (1-x)^4}$$

Rearranging:

$$\frac{x^2(2-x)^2}{(1-x)^4} = \frac{(1.782 \times 10^{-4})(27)(10^2)}{16} = 3.008 \times 10^{-2}$$

Taking the square root of both sides:

$$\frac{x(2-x)}{(1-x)^4} = (3.008 \times 10^{-2})^{\frac{1}{2}} = 0.1734 \qquad\qquad \textbf{(11.25)}$$

This equation can be multiplied out to give an equation of the form $ax^2 + bx + c = 0$ and solved, using the quadratic formula. A simpler method is to use successive approximations. Because ΔG^0 is positive and moderately large we make the initial guess that only a small fraction of the N_2 reacts with H_2. In other words, x is small when compared with 1 or 2. As a first approximation we can thus write (11.25) as

$$\frac{x(2)}{(1)} \approx 0.1734$$

or

$$x \approx 0.0867$$

Feeding this result back into (11.25) enables us to make a second, better approximation.

$$\frac{x(2-0.0867)}{(1-0.0867)^2} \approx 0.1734$$

Solving,

$$x = 0.1734 \frac{(0.9133)^2}{1.9133} = 0.0756$$

A third approximation yields the final result,

$$x = 0.0769$$

The example just calculated was not chosen at random. Because of its economic importance, the equilibrium between nitrogen, hydrogen, and ammonia has been studied experimentally in great detail. The system has been investigated under the precise conditions just calculated, namely 10 atm, 400°C, and an initial mixture of one part N_2 to three parts H_2 (by moles). The experimental value for x is 0.0741. The close agreement with the value we have calculated gives an idea of how reliable thermodynamic tables are in predicting the exact composition of an equilibrium mixture.

We can apply equation (11.14) to equilibria which involve solids and liquids as well as to purely gaseous equilibria. There is one proviso, though. The only mixing which occurs must be in the gas phase. To deal with the thermodynamic consequences of mixing in the solid and liquid phase, i.e., the formation of solutions, is beyond the scope of this text.

An example of a reaction in which only pure condensed phases take part is the equilibrium between carbon dioxide and carbon monoxide gases in contact with solid graphite at 1000°K and 2 atmospheres pressure:

$$C(\text{graphite}) + CO_2(g) \rightleftharpoons 2\,CO(g) \quad (2\ \text{atm},\ 1000°K) \qquad \textbf{(11.26)}$$

If this system is in equilibrium, then the free energy is a minimum. By the same argument as that used above, this can only come about if the free energy of one mole of graphite and one mole of CO_2 gas in the equilibrium mixture will be exactly equal to that of two moles of CO gas. In other words:

$$G(\text{graphite}) + G(CO_2) = 2G(CO) \qquad \textbf{(11.27)}$$

We can write for the two gases concerned:

$$G(CO_2) = G^0(CO_2) + RT \log_e P_{CO_2}$$

and

$$G(CO) = G^0(CO) + RT \log_e P_{CO}$$

while for the solid graphite we can assume that the free energy does not vary measurably with pressure, so that

$$G(\text{graphite}) = G^0(\text{graphite})$$

Feeding these free energy expressions back into (11.27), we find:

$$G^0(\text{graphite}) + G^0(CO_2) + RT \log_e P_{CO_2} = 2G^0(CO) + 2RT \log_e P_{CO}$$

so that

$$2G^0(CO) - G^0(\text{graphite}) - G^0(CO_2) = \Delta G^0$$

$$= -RT \log_e \frac{(P_{CO})^2}{P_{CO_2}}$$

If we write

$$K_p = \frac{(P_{CO})^2}{P_{CO_2}} \qquad\qquad (11.28)$$

then again

$$\Delta G^0 = -RT \log_e K_p$$

where K_p incorporates the partial pressures of all the gases involved in the usual way, but does not feature any term corresponding to the solid. In general, if an equilibrium involves pure liquids and solids as well as gases, only the gases will appear in the expression for K_p. Equation (11.14) is then valid.

Let us now find the equilibrium composition for reaction (11.26). We will calculate how many moles of CO will be formed when one mole of CO_2 is passed over graphite at $1000°K$ and two atmospheres pressure. For this purpose we shall use a value for ΔG^0 of -1.091 kcal.

Inserting this value into

$$\Delta G^0 = -RT \log_e K_p$$

we have

$$-1091 \text{ cal} = -1.987 \text{ cal deg}^{-1} \times 1000 \text{ deg} \times 2.303 \log_{10} K_p$$

so that

$$\log_{10} K_p = 0.2384$$

or

$$K_p = 1.732$$

Suppose that x moles of CO_2 react to form 2x moles of CO at equilibrium. There will then be present in the equilibrium mixture (1-x) moles of CO_2 and 2x moles of CO, giving a total of $(1-x) + 2x = (1 + x)$ moles of gas. By Dalton's law we then have

$$P_{CO_2} = \frac{1-x}{1+x}(2 \text{ atm})$$

and

$$P_{CO} = \frac{2x}{1+x}(2 \text{ atm})$$

Inserting these expressions into equation (11.28), we obtain

$$K_p = \frac{\left(\dfrac{2x}{1+x}\right)^2 (2^2)}{\left(\dfrac{1-x}{1+x}\right)(2)}$$

or

$$1.732 = \frac{8x^2}{(1+x)(1-x)} = \frac{8x^2}{1-x^2}$$

giving

$$x^2 = \frac{1.732}{9.732} = 0.1780$$

so that

$$x = 0.422$$

There will be $(1-x)$ or 0.578 mole of CO_2 and $2x$ or 0.844 mole of CO in the equilibrium mixture.

FREE ENERGY DATA

When dealing with an equilibrium at the standard temperature of $25°C$, it is a relatively simple matter to calculate ΔG^0 (and hence K_p) from tables of ΔG_f^0 values for $25°C$ such as appear in Appendix 3. We have already discussed how to use the tables in Chapter 9.

In order to deal with equilibria at temperatures other than $25°C$, we need more extensive thermodynamic data than Appendix 3 provides. In particular, we need free energy data which cover a range of temperatures. Such information is supplied by the tables given in Appendix 4.

In this appendix, ΔG_f^0 values for a limited number of substances is given at one hundred degree intervals over the range from 300 to $1000°K$. These ΔG_f^0 values can be manipulated in exactly the same way as the $25°C$ values. Note particularly that the ΔG_f^0 value for the stable state of an element is taken as zero at *all* temperatures, and not only at $25°C$.

As an example of the use of these tables, let us find the value of ΔG^0, quoted earlier in the chapter, for the reaction

$$C(graphite) + CO_2(g) \rightleftharpoons 2CO(g)$$

at $1000°K$.

We find from Appendix 4:

Substance	$\Delta G_{f\,1000}^0 (kcal)$
C(graphite)	0.00
$CO_2(g)$	−94.63
CO(g)	−47.86

Thus

$$\Delta G_{1000}^0 = 2(-47.86) - (0.00 - 94.63)\,kcal$$

$$= -1.09\,kcal$$

As a second example of the use of these tables, we will calculate ΔG^0 for the reaction

$$N_2(g) + 3H_2(g) \rightarrow 2NH_3(g)$$

at $400°C = 673.15°K$.

We find from Appendix 4:

| Substance | $\Delta G_f^0 (kcal)$ | |
	$600^\circ K$	$700^\circ K$
$N_2(g)$	0.00	0.00
$H_2(g)$	0.00	0.00
$NH_3(g)$	+3.78	+6.51

Thus at $600^\circ K$,

$$\Delta G_{600}^\circ = 2\,(3.78) - 0.00 - 3\,(0.00) = 7.56\,kcal$$

while at $700^\circ K$,

$$\Delta G_{700}^\circ = 2\,(6.51) - 0.00 - 3\,(0.00) = 13.02\,kcal$$

The value at $700^\circ K$ is 5.46 kcal higher than that at $600^\circ K$. Assuming a linear variation, the value at $673.15^\circ K$ is then 5.46 (73.15/100) = 3.99 kcal higher than at $600^\circ K$; i.e.,

$$\Delta G_{673}^\circ = 7.56 + 3.99 = 11.55\,kcal$$

This is the value of ΔG^0 used previously to calculate the extent of this reaction at this temperature.

The use of the free energy data in Appendix 4, as just described, enables us to find ΔG^0 for a large number of chemical reactions over a wide range of temperatures. Even though the tables are restricted to a limited number of substances, they enable us to make accurate predictions about the equilibrium states of over a thousand reactions at any temperature between $300^\circ K$ and $1000^\circ K$.

It is useful to know where further free energy data, similar to that in Appendix 4, can be found. The JANAF tables, published by the Dow Chemical Co., include over a thousand compounds, mainly inorganic gases and solids, and cover the range from 0° to $6000^\circ K$. Values for gaseous organic carbon compounds can be found in "The Chemical Thermodynamics of Organic Compounds" by Stull, Westrum, and Sinke, published by John Wiley and Sons, Inc. There is a third book of thermodynamic data which is also very valuable. This is the Circular No. 500 of the National Bureau of Standards, which contains very extensive tables, similar to those in Appendix 3, of ΔH^0, ΔG^0, and S^0 values. The values given in this circular are restricted to $25^\circ C$, however.

FREE ENERGY AND MOLECULAR BEHAVIOR

When using tables of free energies to calculate values for ΔG^0 and hence equilibrium concentrations, students often develop misgivings about the whole calculation process. They get the feeling that such calculations, though useful, are mere juggling with numbers. There seems to be very little contact with "real" chemistry.

This attitude is difficult to dispel, largely because it has an element of truth in it. Although a value of ΔG^0 for a reaction enables us to predict exactly where the equilibrium position will be at a given temperature and pressure, it tells us *nothing,* by itself, about the *reasons* for this on a *molecular level.* It is not until we break down the value of ΔG^0 into its constituent ΔH^0 and $T\Delta S^0$ terms that we can begin to understand the equilibrium in molecular terms.

Suppose, for instance, we have a hypothetical reaction

$$A(g) + B(g) \rightarrow C(g) \tag{11.29}$$

occurring at $25°C$, where A, B, and C are unspecified molecules. If we are told that ΔG^0 for the reaction is -16.6 kcal, this information enables us to calculate how much C will be formed when a given number of moles of A and B are mixed at a given pressure. Indeed, we can discern at a glance that ΔG^0 is sufficiently negative for the reaction to go almost to completion under most circumstances at this temperature. Nevertheless, this value of ΔG^0 gives us *no* information about the molecules A, B, and C, or why, on a molecular level, the reaction goes virtually to completion.

On the other hand, if we are also informed that $\Delta H^0 = -30$ kcal and $\Delta S^0 = -45$ cal deg^{-1} for this reaction, this new knowledge *does* tell us something about A, B, and C on a molecular level. Of course, it doesn't tell us *everything* about these molecules, but it tells us everything of importance in determining the position of chemical equilibrium. It informs us about the relative energies and the relative spacings of the energy levels involved. Thus, the fact that $\Delta H^0 = -30$ kcal for this reaction means that it requires energy to decompose a molecule of C in its ground state into a molecule of A and a molecule of B each in their ground states. In other words, the value of ΔH^0 tells us that the energy levels corresponding to the left-hand side of reaction (11.29) are on the whole very much higher in energy than those on the right-hand side.

In contrast to the value of ΔH^0, which tells us about the *relative energies* of levels, the value of ΔS^0 informs us about the *relative spacings* of levels. The fact that ΔS^0 for reaction (11.29) is negative means that the standard entropy of the substances on the left-hand side of the equation is much larger than that of the substances on the right. In other words, when a molecule of C decomposes into a molecule of A plus a molecule of B, the number of energy levels available for occupation is enormously increased. The reason for this is that the constraint on the movement of the nuclei in the two independent fragments A and B is much less than the constraint on the combined entity C in which the two fragments are forced to move around together.

It is only when we have looked at the reaction in this way and seen which is the low-energy side and which is the high entropy side that we can begin to understand in molecular terms why the reaction of A with B goes virtually to completion at 25°C, or, what is really the same thing, why compound C decomposes into A + B at this temperature to only a negligible extent. It is the *very large difference in energy between the ground states* which is the main factor in determining the position of the equilibrium. The temperature is simply not high enough for any of the much more numerous levels corresponding to the species A + B to have even the suggestion of a population. The fact that there are a very large number of such levels is of little significance at 25°C. Only at higher temperatures, when these levels begin to acquire appreciable populations, does the fact that there are a large number of them become important.

Another way of appreciating that ΔG^0 values are not just magic numbers derived from tables, but that they have some basis on a molecular level, is by considering some examples of chemical equilibria which are simple enough to allow us to estimate an approximate value for ΔG^0 from a minimum of information. The first of these examples is the reaction

$$CO(g) + 3H_2(g) \rightarrow CH_4(g) + H_2O(g) \quad (T°K, 1 \text{ atm}) \qquad \textbf{(11.30)}$$

We must first estimate ΔH^0 for this reaction. We do this by using the bond energy values of Table 5-3. In the usual fashion we find that

$$\Delta H^0 \approx D_{C-O} + 3D_{H-H} - 4D_{C-H} - 2D_{O-H}$$

$$= (256 + 3 \times 104 - 4 \times 99 - 2 \times 111) \text{ kcal}$$

$$= (568 - 618) \text{ kcal} = -50 \text{ kcal}$$

The reaction is exothermic because more bonds are made than are broken, even though one of the bonds which is broken is the conspicuously strong C≡O bond.

We must next estimate ΔS^0. It is possible to find an accurate value for this quantity at 25°C from tables. However, since the purpose of this exercise is to use as little tabulated information as possible, let us try to make an educated guess. Most diatomic molecules, and simple polyatomic molecules containing hydrogen, have S^0_{298} values within the range from 45 to 55 cal deg^{-1} mole^{-1}. Only heavy molecules such as I_2 lie outside this range. Accordingly, we shall take a mean value of 50 cal deg^{-1} mole^{-1} as the S^0_{298} value for the gases CO, CH_4, and H_2O. The value of S^0_{298} for H_2 is distinctly smaller than this, because H_2 is so much lighter and smaller than other diatomic molecules. The tabulated value is 31.2 cal deg^{-1} mole^{-1}. Here we shall round it off to a value of 30 cal deg^{-1} mole^{-1}.

Using these approximate S^0 values we estimate that

$$\Delta S^0 = (2 \times 50 - 50 - 3 \times 30)\ \text{cal deg}^{-1}$$

$$= -40\ \text{cal deg}^{-1}$$

(the correct value is actually -51.2 cal deg^{-1}).

Combining the approximate values of ΔH^0 and ΔS^0 we can now estimate ΔG^0 at 25°C:

$$\Delta G^0 = \Delta H^0 - T\Delta S^0$$

$$\approx -50 \times 10^3\ \text{cal} + 298\ \text{deg} \times 40\ \text{cal deg}^{-1}$$

$$= -50{,}000 + 11{,}920\ \text{cal}$$

$$= -38\ \text{kcal}$$

The correct value of ΔG^0 is -33.9 kcal, so our approximation is in error by about 4 kcal. Even if our estimate of ΔG^0 were much less reliable than this, we could still conclude that ΔG^0 for this reaction is much more negative than the -10 kcal value previously suggested as a rough criterion for a reaction going to completion at 25°C. We would thus expect the reaction of hydrogen with carbon monoxide to go almost to completion at this temperature.

In practice this reaction, like many reactions involving H_2 at 25°C, proceeds so slowly that we cannot observe it occurring. It can only be made to go at a measurable rate by using a catalyst and a somewhat higher temperature. At 250°C, for instance, with a nickel catalyst, the reaction is observed to go to completion. We can see that ΔG^0 is still a large negative quantity at this temperature by using the same ΔH^0 and ΔS^0 values estimated above. We have:

$$\Delta G^0 = \Delta H^0 - T\Delta S^0$$

$$\approx -50 \times 10^3\ \text{cal} + 523\ \text{deg} \times 40\ \text{cal deg}^{-1}$$

$$= -50{,}000 + 20{,}920\ \text{cal}$$

$$= -29\ \text{kcal}$$

As we shall see a little later on, a value of ΔG^0 of -29 kcal corresponds to a reaction that goes virtually to completion at 250°C.

As a second example of the calculation of free energy changes from a minimum of data, let us discuss not a single reaction but rather a whole set of reactions of the same type. We will consider the addition of H_2 to a C$=$C bond, a process usually called *hydrogenation,* and the reverse of this process, usually called *dehydrogenation.* The simplest example of hydrogenation is the reaction

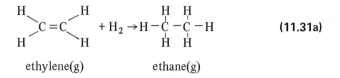

$$\text{ethylene(g)} \qquad \text{ethane(g)} \qquad \textbf{(11.31a)}$$

Other examples are

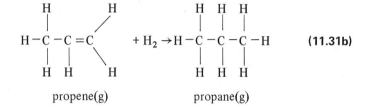

$$\text{propene(g)} \qquad \text{propane(g)} \qquad \textbf{(11.31b)}$$

and

$$C_6H_5 - \overset{\displaystyle H}{\underset{\displaystyle |}{C}} = C \overset{\displaystyle H}{\underset{\displaystyle H}{\diagup\diagdown}} \quad + H_2 \rightarrow C_6H_5 - \overset{\displaystyle H}{\underset{\displaystyle |}{\underset{H}{C}}} - \overset{\displaystyle H}{\underset{\displaystyle |}{\underset{H}{C}}} - H \qquad \textbf{(11.31c)}$$

$$\text{styrene(g)} \qquad\qquad \text{ethylbenzene(g)}$$

In each of these reactions one H–H bond is broken and two C–H bonds are made. In addition, a carbon-carbon double bond is replaced by a carbon-carbon single bond. ΔH^0 should then be approximately the same for all three reactions.

$$\Delta H^0 \approx D_{\text{H–H}} + D_{\text{C=C}} - 2D_{\text{C–H}} - D_{\text{C–C}}$$

$$= (104 + 147 - 2 \times 99 - 83) \text{ kcal}$$

$$= (251 - 281) \text{ kcal}$$

$$= -30 \text{ kcal}$$

Notice that these reactions are exothermic because more bonds are made than are broken.

ΔS^0 should also be approximately the same for all three reactions. In each case the H_2 molecule attaches itself to a much heavier molecule. The addition of two light H atoms to a heavy molecule will increase its entropy to only a small extent. In particular, there will be very little change in the translational and rotational contributions to the entropy. We thus expect the standard molar entropy of ethane to be only slightly larger than that of ethylene, that of propane to be only slightly larger than that of propene, and so on.

Because the entropy of the product and one of the reactants will be almost the same in each case, ΔS^0 should be approximately equal to $-S^0(H_2)$ for all three reactions. The standard molar entropy of H_2 is 31.2 cal deg^{-1} $mole^{-1}$ at $25°C$. As in our previous example, we shall round this figure off to 30 cal deg^{-1} $mole^{-1}$.

In Table 11-1, the accepted values of ΔH^0 and ΔS^0 for these three reactions are given. As can be seen, the approximations that $\Delta H^0 = -30$ kcal and $\Delta S^0 = -30$ cal deg^{-1} are good ones.

Using these approximate values for ΔH^0 and ΔS^0, we can now calculate ΔG^0 for a hydrogenation reaction at $25°C$. We find that:

$$\Delta G^0 = \Delta H^0 - T\Delta S^0$$

$$= (-30 + 298 \times 30 \times 10^{-3}) \text{ kcal}$$

$$= (-30 + 8.9) \text{ kcal} = -21.1 \text{ kcal}$$

In other words, we can expect any hydrogenation reaction which is essentially similar to the three considered here to go to completion at $25°C$. This expectation is realized in practice. The catalytic hydrogenation of carbon compounds at room temperature is a standard method of determining the number of carbon-carbon double bonds in the molecule.

Since the low energy and high entropy sides of hydrogenation reactions are different, we expect such reactions to change direction at high temperatures; i.e., we expect dehydrogenation rather than hydrogenation to occur when T is raised. The approximate temperature at which this change-over occurs will be that at which ΔG^0 becomes zero. We can easily estimate this temperature if we use the

Table 11-1. ΔH^0 and ΔS^0 for Several Hydrogenation Reactions at $25°C$.

Reaction	$\Delta H°$ (kcal)	$\Delta S°$ (cal deg^{-1})
$C_2H_4 + H_2 \rightarrow C_2H_6$	-32.74	-28.81
$C_3H_6 + H_2 \rightarrow C_3H_8$	-29.70	-30.51
$C_8H_8 + H_2 \rightarrow C_8H_{10}$	-28.10	-27.54

rough values of ΔH^0 and ΔS^0 obtained above. If ΔG^0 is zero, then

$$\Delta H^{\circ} = T\Delta S^0$$

$$30 \text{ kcal} = T \times 30 \text{ cal deg}^{-1}$$

$$T = 1000^{\circ}\text{K}$$

$$= 727^{\circ}\text{C}$$

In point of fact styrene, which is of considerable commercial importance in the manufacture of plastic foam, is made industrially by the dehydrogenation of ethylbenzene, that is, by the reverse of reaction (11.31c). The reaction is carried out at a temperature of 650°C, a little lower than our estimate.

The ability to predict an approximate value for ΔG^0 with a minimum of information not only is useful in helping us to see that free energy differences depend ultimately on molecules; it also allows us to discuss equilibria involving compounds for which there is little or no thermodynamic data available. It even allows us to discuss the thermodynamics of *non-existent* compounds. Sometimes such a discussion enables us to understand why the compound does not exist. A case in point is the compound sulfur monoxide, SO.

It is not quite true to say that sulfur monoxide does not exist. It can exist in low concentrations at high temperatures and its spectrum has been investigated. But no one has yet isolated it in the laboratory. On the face of it, this is surprising, since it agrees with all the elementary rules of the octet theory. Both sulfur and oxygen have a valence of two, since they need two electrons to make up their octet. We can readily write a Lewis formula for SO analogous to that for O_2:

$$\ddot{\text{S}} :: \ddot{\text{O}} \quad \text{and} \quad \ddot{\text{O}} :: \ddot{\text{O}}$$

The non-existence of SO does not come about because of any thermodynamic obstacles to preparing it from its elements. ΔG_f^0 can be estimated for SO at a variety of temperatures. Such estimates vary from -1.5 kcal at 100°K to -16 kcal at 1000°K, and except at very low temperatures they are always negative. Without further information, one would predict the ready formation of SO from its elements. However, sulfur monoxide is unstable because it can convert to another oxide, the dioxide, according to the equation

$$4 \text{ SO(g)} \rightarrow 2 \text{ SO}_2(g) + \text{S}_2(g) \qquad \textbf{(11.32)}$$

The occurrence of this reaction involves the breaking of four S=O bonds in sulfur monoxide and the making of four similar S=O bonds in the two SO_2 molecules. In addition a fifth bond, the S=S bond of S_2, must also be made. If the S=O bond energies in the two molecules are much the same, then ΔH for reaction (11.32) must be negative by an amount close to the bond energy of the additional S=S bond. From Table 5-3, this bond energy is 100 kcal mole^{-1}. We thus estimate ΔH^0 as -100 kcal at any temperature.

We can also obtain a rough estimate for the entropy change by assuming that the two diatomic gases have S^0 values of 50 cal deg^{-1} mole^{-1}, while that of the triatomic gas is a little higher, say 55 cal deg^{-1} mole^{-1}. We then estimate

$$\Delta S^0 = (2 \times 55 + 50 - 4 \times 50) \text{ cal deg}^{-1}$$

$$= -40 \text{ cal deg}^{-1}$$

Substituting these two approximate values into the expression for ΔG^0, we estimate that at 25°C,

$$\Delta G^0 = \Delta H^0 - T\Delta S^0$$

$$= -100 \times 10^3 \text{ cal} + 298 \text{ deg} \times 40 \text{ cal deg}^{-1}$$

$$= -100,000 + 11,920 \text{ cal}$$

$$= -88 \text{ kcal}$$

There can be little doubt, no matter how gross our approximations may have been, that reaction (11.32) will go to completion at 25°C!

ΔG^0 AND THE EXTENT OF A REACTION

In Chapter 9 it was suggested as a rough rule of thumb for a reaction at 25°C that if ΔG^0 is more negative than -10 kcal the reaction goes virtually to completion, while if ΔG^0 is more positive than $+10$ kcal, scarcely any reaction occurs. Useful as such a rule is, it has several drawbacks. The phrase "virtually to completion" is not very precise. The rule neglects the fact that the extent of reaction depends on the reaction type as well as on the value of ΔG^0. And finally, it tells us nothing about reactions at temperatures different from 25°C.

Table 11-2 helps us to get over these difficulties. In this table the extent* of a reaction is tabulated against ΔG^0_{298} for various reaction types. From this table we can predict the approximate extent of a reaction from a given value of ΔG^0

---

*The extent of a reaction is defined in Chapter 9 (p. 198).

Table 11-2.

Extent of Reaction as a Function of $\Delta G^0/T$ and ΔG^0_{298} for Various Reaction Types at One Atmosphere Pressure.

	ΔG^0 at 25°C (kcal)														
	+11.9	+8.9	+6.0	+3.0	+1.8	+1.2	+0.6	0.0	-0.6	-1.2	-1.8	-3.0	-6.0	-8.9	-11.9
A ⇌ B	↓	↓	↓	0.006	0.047	0.118	0.268	0.500	0.732	0.882	0.953	0.994	↑	↑	↑
A ⇌ 2B	↓	↓	0.003	0.040	0.110	0.180	0.289	0.447	0.637	0.807	0.915	0.987	↑	↑	↑
A+B ⇌ C	↓	↓	↓	0.003	0.024	0.061	0.144	0.293	0.483	0.657	0.784	0.975	0.993	0.999	↑
A+B ⇌ 2C	↓	↓	0.003	0.039	0.100	0.155	0.232	0.333	0.453	0.578	0.693	0.861	0.987	0.999	↑
A+B ⇌ C+D	↓	0.001	0.006	0.075	0.181	0.268	0.377	0.500	0.623	0.732	0.819	0.925	0.994	0.999	↑
A+2B ⇌ C+D	↓	0.001	0.007	0.083	0.192	0.274	0.372	0.477	0.580	0.672	0.751	0.863	0.973	0.995	0.999
A+2B ⇌ 2C	↓	↓	0.004	0.044	0.109	0.166	0.241	0.333	0.437	0.458	0.639	0.792	0.957	0.992	0.999
A+3B ⇌ C+D	↓	0.001	0.008	0.091	0.203	0.284	0.376	0.473	0.566	0.649	0.720	0.826	0.950	0.986	0.996
A+3B ⇌ 2C	↓	↓	0.004	0.049	0.119	0.176	0.252	0.340	0.436	0.531	0.619	0.758	0.929	0.980	0.994
	+40	+30	+20	+10	+6	+4	+2	0.0	-2	-4	-6	-10	-20	-30	-40
	$\Delta G^0/T = R \log_e K$ (cal deg^{-1})														

at $25°C$. As an example of the use of this table, let us take the reaction

$$Br_2 + 2 NO \rightarrow 2 NOBr \quad (25°C, 1 \text{ atm})$$

The value of ΔG^0 for this reaction at $25°C$ can be found from the tables in Appendix 3. It is -1.98 kcal. Because this reaction is of the type $A+2B \rightleftharpoons 2C$, we consult the row corresponding to this type in Table 11-2. There we find that an extent of 0.639 corresponds to a ΔG^0 value of -1.8 kcal, while an extent of 0.792 corresponds to a ΔG^0 value of -3.0 kcal. The true extent must lie between these two values and is thus close to 0.7. In other words, if 1 mole of Br_2 gas and two moles of NO gas are allowed to attain equilibrium at $25°C$ and 1 atmosphere pressure, approximately 0.7 mole of Br_2 will be destroyed and $2 \times 0.7 = 1.4$ moles of NOBr will be formed.

A close inspection of Table 11-2 enables us to see under what conditions the ± 10 kcal rule can be expected to hold. We find from the table that if ΔG^0 has a value more positive that $+8.9$ kcal, none of the reaction types shown will proceed to an extent greater than 0.001. For negative values of ΔG^0 much the same holds true. If ΔG^0 is more negative than -8.9 kcal, most reactions are better than 99 per cent complete. For ΔG^0 more negative than -11.9 kcal, this figure rises to almost 99.9 per cent for all the reactions shown.

These figures show that the limits ± 10 kcal correspond quite well to an extent of reaction between 0.1 per cent and 99.9 per cent, at least for the reaction types shown. If more than six or seven molecules are involved in the reaction, the limits must be extended beyond 10 kcal. We should also remember that these figures apply only to $25°C$, to one atmosphere pressure, and to stoichiometric amounts of reactants. In the example just quoted we were able to use Table 11-2 to find the approximate extent of a reaction when one mole of Br_2 was mixed with *two* moles of NO at *one* atmosphere pressure. We would not have been able to use the table to predict what would happen if one mole of Br_2 had been mixed with 5.32 moles of NO at 1.87 atmospheres pressure.

Table 11-2 can also be used for temperatures other than $25°C$. In order to do this, the row corresponding to $\Delta G^0/T$ at the bottom of the table must be consulted instead of the top row of ΔG^0_{298} values. Suppose, for instance, we have a reaction of the type $A+2B \rightleftharpoons 2C$ at $1000°K$ for which ΔG^0 has the value -10 kcal. The value of $\Delta G^0/T$ is first calculated and found to be -10 cal $\deg^{-1}$. From the table we then find that the extent corresponding to this value of ΔG^0 for this reaction type is 0.792. Thus, if one mole of A and two moles of B are allowed to react at this temperature and pressure, 0.792 mole of A and 2×0.792 mole of B will react to produce $2 \times 0.792 = 1.584$ moles of C.

It is possible to develop from Table 11-2 a rule equivalent to the ± 10 kcal rule, but expressed in terms of $\Delta G^0/T$ and thus valid at any temperature. As before, we consider a reaction to have scarcely occurred at all if the extent is less than 0.001, and to have gone virtually to completion if the extent is greater than 0.999. For the reaction types given in the table we find that these two limits correspond to $\Delta G^0/T = +30$ cal deg^{-1} at the one end and to $\Delta G^0/T = -40$ cal deg^{-1} at the other. An average of these two values, namely 35 cal deg^{-1}, then gives a convenient, if rough, guide as to when a reaction will scarcely occur or go virtually to completion. If $\Delta G^0/T$ is more positive than $+35$ cal deg^{-1}, the reaction will occur to an extent less than about 0.1 per cent. If $\Delta G^0/T$ is more negative than -35 cal deg^{-1}, the reaction will occur to an extent greater than about 99.9 per cent.

As an example of the use of this rule, let us return to the reaction between carbon monoxide and hydrogen:

$$CO(g) + 3H_2(g) \rightarrow CH_4(g) + H_2O(g)$$

In considering this reaction previously, we made a rough estimate of ΔG^0 for this reaction at a temperature of $250^\circ C$ ($523^\circ K$) and found it to be -29 kcal. This corresponds to a value of -55.4 cal deg^{-1} for $\Delta G^0/T$, well below the -35 cal deg^{-1} limit. We can safely conclude that if one mole of CO and three moles of H_2 are passed over a catalyst at one atmosphere pressure at $250^\circ C$, the reaction will go to completion.

At various points in this chapter we have estimated an approximate value for ΔG^0. It is both interesting and useful to know just how good or how bad such an approximate value of ΔG^0 is for predicting the equilibrium composition. The use of Table 11-2 enables us to do this rather well.

Consider, for instance, the reaction

$$N_2O_4(g) \rightleftharpoons 2\,NO_2(g) \quad (1 \text{ atm}, 25^\circ C)$$

ΔG^0 for this reaction is reliably estimated as $+1.15$ kcal and the extent of the reaction at equilibrium has been calculated above and shown to be 0.186. Suppose now that we were uncertain as to the value of ΔG^0 by some 5 kcal in either direction so that we knew only that ΔG^0 lay somewhere between $+6.15$ kcal and -3.85 kcal. From Table 11-2 we find that $\Delta G^0 = +6.15$ kcal corresponds to an extent of about 0.003, while $\Delta G^0 = -3.85$ corresponds to an extent of more than 0.987. In other words, if there is an uncertainty of ± 5 kcal in the value of ΔG^0, we are unable to do better than say that N_2O_4 dissociates somewhere between 0.3 per cent and 99 per cent.

We next consider a reaction for which ΔG^0 has a large rather than a small value. Previously, it was estimated that ΔG^0 for the reaction

$$H_2C = CH_2(g) + H_2(g) \rightarrow H_3C - CH_3 \quad (25^\circ C)$$

is approximately -21 kcal. Let us suppose, as before, that there is an uncertainty of ± 5 kcal in this value of ΔG^0, so that it may have any value lying between the limits -16 kcal and -26 kcal. From Table 11-2 we find that for a reaction of this type $(A + B \rightarrow C)$ the extent of the reaction is greater than 99.9 per cent over the whole of this range of values for ΔG^0. Despite an uncertainty of ± 5 kcal in the free energy difference, we can tell that the reaction will go to completion.

The two examples just quoted exemplify the two extreme results of an uncertainty in the value of ΔG^0. If ΔG^0 is close to zero, we need to know its value very precisely in order to be able to predict the equilibrium composition with any accuracy. An uncertainty of ±5 kcal is much too large to permit a reliable estimate of the extent of a reaction to be made. On the other hand, if ΔG^0 is very large, whether positive or negative, an uncertainty of ±5 kcal is of no consequence. The reaction scarcely occurs at all, or it goes virtually to completion, and a variation of 5 kcal in the value of ΔG^0 makes no difference to such a result.

CONCLUSION

In this chapter we have finally succeeded in making the treatment of chemical equilibrium *quantitative*. Now that we have established relationship (11.14),

$$\Delta G^0 = -RT \log_e K_p$$

we are able to translate thermodynamic quantities into an equilibrium constant and hence predict the composition of an equilibrium mixture. This represents a major advance over the low energy–high entropy rule. All that this rule could tell us was whether reactants or products would be "favored" at high or low temperatures. Equation (11.14), on the other hand, allows us to predict the exact extent of a reaction at a given temperature.

The efficacy of equation (11.14) depends, of course, on how accurate a value of ΔG^0 we are able to obtain. This is a subject which we have investigated at some length. In the first place, we saw how to manipulate free energy tables so as to be able to obtain an accurate value of ΔG^0 at any temperature. We also investigated under what circumstances an inaccurate value of ΔG^0 could be useful. As a result, we obtained some insight into what a value of ΔG^0 means in terms of how far a reaction will go in one direction or the other at any temperature.

This treatment of the relationship between free energy and equilibrium brings our present chapter to an end, and with it our account of thermodynamics.

Basically, this book has dealt with one subject only — chemical equilibrium. We have looked at this subject from three very different but mutually complementary points of view. In Chapter 1, we regarded equilibrium in a rather abstract way, in terms of probabilities. This led us to characterize an equilibrium state as the *most probable state*. Our second approach (Chapters 2 to 4) was in terms of molecules, or, more specifically, in terms of molecular energy levels. Viewed in this way, chemical equilibrium was seen to be the outcome of a *competition between alternative sets of energy levels* for population. The third point of view adopted (Chapters 5 to 11) was one which used thermodynamic functions. This last approach finally led us to regard chemical equilibrium as corresponding to a position of *minimum free energy.*

Each of these three approaches has its use, and our appreciation of chemical equilibrium is incomplete without all three. By looking at the subject in terms of *probability,* for instance, we see why chemical equilibrium occurs in the first place. The probability of anything else happening, such as the reaction reversing itself or a measurable fluctuation occurring, is simply too small.

Looking at chemical equilibrium in terms of free energy, by contrast, contributes very little to our understanding on a *molecular* basis. What this approach *does* is to enable us to handle the subject on a *numerical* basis. We eventually find ourselves in the convenient situation of being able to calculate the equilibrium constant and hence the equilibrium composition for a very large number of reactions from a relatively small table of numbers.

Finally, to look at chemical equilibrium in terms of *energy levels* is to see what is at stake in terms of molecules. Such an insight prevents us from forgetting that the direction of a chemical reaction or the position of a chemical equilibrium depends ultimately on the properties of molecules and the structure of matter.

PROBLEMS

1. Write out expressions for K_p in terms of partial pressures for the following reactions.
   a. $H_2(g) + I_2(g) \rightleftharpoons 2HI(g)$
   b. $H_2(g) + I_2(s) \rightleftharpoons 2HI(g)$
   c. $CuO(s) + H_2(g) \rightarrow Cu(s) + H_2O(g)$
   d. $CS_2(l) \rightarrow CS_2(g)$
   e. $CaCO_3(s) \rightarrow CaO(s) + CO_2(g)$

2. Calculate the value of ΔG^0 at 25°C for the following reactions from the ΔG_f^0 values given in Appendix 3. From these, deduce the extent of each of the reactions at 25°C and 1 atmosphere total pressure when stoichiometric amounts of the reactants are mixed. (Use Table 11-2.)
   a. $CO_2(g) + NO(g) \rightarrow CO(g) + NO_2(g)$
   b. $I_2(s) + Cl_2(g) \rightarrow 2ICl(g)$
   c. $2Ag(s) + H_2S(g) \rightarrow Ag_2S(s) + H_2(g)$
   d. $2H_2S(g) + 3O_2(g) \rightarrow 2H_2O(g) + 2SO_2(g)$
   e. $N_2(g) + O_2(g) \rightarrow 2NO(g)$
   f. $2F_2(g) + O_2(g) \rightarrow 2F_2O(g)$
   g. $2Cl_2(g) + O_2(g) \rightarrow 2Cl_2O(g)$
   h. $PbO(red) \rightarrow PbO(yellow)$
   i. $CH_4(g) + CO_2(g) \rightarrow CH_3OH(g) + CO(g)$
   j. $N_2F_4(g) \rightarrow 2NF_2(g)$
   k. $2NO(g) + Br_2(g) \rightarrow 2NOBr(g)$
   l. $2Ag(s) + I_2(s) \rightarrow 2AgI(s)$
   m. $SO_3(g) + H_2(g) \rightarrow SO_2(g) + H_2O(g)$

3. A and B are two *optical* isomers. Calculate ΔH^0, ΔS^0, ΔG^0 and K_p for the reaction

$$A(g) \rightarrow B(g)$$

at 25°C.

4. From ΔG_f^0 values given in Appendix 3, calculate the vapor pressure of the two liquids CCl_4 and CS_2 at 25°C.

5. Use the tables given in Appendix 3 to find ΔH and ΔS for the process

$$CCl_4(l) \rightarrow CCl_4(g) \quad (25°C, 1 \text{ atm})$$

Assuming that neither ΔH nor ΔS varies with temperature, estimate the boiling point of CCl_4. In a similar way, estimate the boiling point of CS_2. The observed values of these boiling points are 76.8°C (CCl_4) and 46.3°C (CS_2).

6. Explain why the value of ΔG_f^0 for $I_2(g)$ given in Appendix 4 is not zero at all temperatures.

7. Use the ΔG_f^0 values in Appendix 4 to find ΔG^0 for the process

$$Na(s) \rightarrow Na(l)$$

at 300°K and 400°K. Then estimate the melting point of sodium. (The observed melting point is 97.5°C.)

8. Careful measurements on the pressure of a mixture of a known amount of NO and Cl_2 gases in a given volume allow K_p for the reaction

$$2 \, NOCl(g) \rightarrow 2NO(g) + Cl_2(g)$$

to be measured over the temperature range from 350 to 500°K. At 422.6°K, the value of K_p obtained in this way is 6.59×10^{-4} atm. Calculate ΔG^0 for the reaction at 422.6°K from this value of K_p and compare it with the value of ΔG_f^0 obtained from the tables in Appendix 4.

9. Gaseous iodine dehydrogenates isobutane to yield isobutene and hydrogen iodide according to the equation

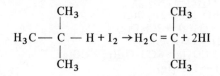

At 582.7°K the following partial pressures were measured for the

equilibrium mixture:

Substance	Partial Pressure (mm)
$i-C_4H_{10}$	163.1
I_2	0.26
$i-C_4H_8$	5.92
HI	11.8

Calculate K_p for this reaction at this temperature. The value of ΔG^0 for this reaction at $582.7°K$ derived from tables is $+4.33$ kcal. Calculate K_p from this value of ΔG^0 and compare it to the experimental value.

10. In an equilibrium mixture of H_2, I_2, and HI gases at $698.2°K$, the following partial pressures were experimentally determined.

Substance	Partial Pressure (atm)
H_2	0.1664
I_2	0.0977
HI	0.9437

Use these values to calculate K_p and hence ΔG^0 for the reaction

$$H_2(g) + I_2(g) \rightarrow 2HI(g) \quad (698.2°K)$$

Compare this value of ΔG_f^0 with that obtained from the ΔG_f^0 tables in Appendix 4.

11. The following partial pressures were determined experimentally for an equilibrium mixture of phosgene, carbon monoxide and chlorine gases at $394.8°C$.

Gas	Partial Pressure (mm)
$COCl_2$	253.9
CO	88.1
Cl_2	97.5

Use these data to calculate K_p for the reaction

$$COCl_2(g) \rightarrow CO(g) + Cl_2(g)$$

at $394.8°C$. Compare this value to that obtained from ΔG_f^0 values in Appendix 4.

12. From tables it is found that the value of ΔG^0 for the isomerization reaction

$$CH_3 - CH_2 - CH_2 - CH_2 - CH_3 \rightarrow \overset{\overset{\displaystyle CH_3}{\displaystyle |}}{CH_3 - CH - CH_2 - CH_3}$$

   n-pentane   isopentane

is 0.67 kcal at $1000°K$. Calculate K_p for this reaction at this temperature. Then find the percentage of each isomer which will be present in an equilibrium mixture of the two at this temperature.

13. At $350°K$, ΔG^0 for the reaction

$$N_2O_4(g) \rightleftharpoons 2NO_2(g)$$

has the value -1.02 kcal. Calculate the number of moles of NO_2 formed at equilibrium when one mole of N_2O_4 is allowed to decompose at this temperature and 1 atmosphere pressure. Next calculate the number of moles formed at 10 atmospheres pressure and the same temperature. Does an increase in pressure result in the production of more NO_2 or less?

14. When one mole of N_2 and three moles of H_2 are mixed at $450°C$ and a total pressure of 10 atmospheres in the presence of a catalyst, 0.0797 mole of NH_3 is formed in the resultant equilibrium mixture.

   Using the ΔG_f^0 values in Appendix 4, deduce the value of ΔG^0 for the reaction

$$N_2(g) + 3H_2(g) \rightarrow 2NH_3(g)$$

at $450°C$, and hence the value of K_p at the same temperature. Use this value of K_p to calculate how many moles of NH_3 will be formed at this temperature and a pressure of 10 atmospheres when one mole of N_2 and three moles of H_2 are mixed. Compare this calculated value with the experimental value given in the previous paragraph.

15. From Appendix 3, find ΔH^0 and ΔS^0 for the reaction

$$N_2(g) + 3H_2(g) \rightarrow 2NH_3(g)$$

at $25°C$. Next calculate ΔG^0 for this reaction at $400°C$, assuming that ΔH^0 and ΔS^0 do not vary with temperature. Compare this value to the more accurate value, 11.55 kcal, used to derive equation (11.24). Next calculate the number of moles of NH_3 formed when one mole of N_2 and three moles of H_2 are allowed to react at $400°C$ and 1 atmosphere pressure using the approximate value of ΔG^0 just calculated. Compare the value obtained to the experimental value of 0.148 mole of NH_3.

16. From the tables of ΔG_f^0 values given in Appendix 4, deduce ΔG^0 and K_p for the reaction

$$N_2O_4(g) \rightleftharpoons 2NO_2(g)$$

at $45°C$. Then calculate how many moles of N_2O_4 will dissociate when one mole is heated to this temperature at a total pressure of 0.2662 atmospheres. Compare the value obtained in this way to the experimentally observed value of 0.624 mole.

17. Experimentally, it is found that when one mole of PCl_5 is heated to 502.2°K at a pressure of 741.7 mm of mercury, 0.608 mole dissociates to PCl_3 and Cl_2 according to the equation

$$PCl_5(g) \rightarrow PCl_3(g) + Cl_2(g)$$

Use the free energy data given in Appendix 4 to deduce ΔG^0 and K_p for this temperature. From this value of K_p, calculate how many moles of PCl_5 will dissociate under the above conditions and compare with the experimental result.

18. Without consulting Appendix 3 or Appendix 4, estimate ΔH^0 and ΔS^0 for the gaseous reaction

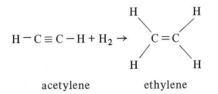

From these values, calculate an approximate value for ΔG^0 at 25°C, and the approximate temperature at which $\Delta G^0 = 0$ for this reaction.

19. Use the ΔG_f^0 values in Appendix 3 to calculate an accurate value of ΔG_{298}^0 for the reaction in the previous example.

20. Pure CO_2 gas is passed slowly over heated graphite and the following reaction occurs:

$$C(graphite) + CO_2(g) \rightarrow 2CO(g)$$

If the emergent gas at one atmosphere pressure contains 50% CO by volume, estimate the temperature of the graphite. Assume equilibrium conditions and use Appendix 4.

21. Two isomers, A and B, are in equilibrium with each other at 500°K. The equilibrium mixture consists of 70 per cent A and 30 per cent B. Find ΔG^0 for the reaction

$$A(g) \rightarrow B(g)$$

at this temperature.

22. At moderately high temperatures, solid calcium carbonate dissociates according to the equation

$$CaCO_3(s) \rightarrow CaO(s) + CO_2(g)$$

The equilibrium pressure of the CO_2 at various temperatures has been measured with the following results:

Temperature (°C)	P_{CO_2} (atm)
500	9.3×10^{-5}
700	2.92×10^{-2}
897	1.000
1100	11.50

Use these pressures to calculate ΔG^0 for the reaction at the four temperatures. Plot these ΔG^0 values against the absolute temperature. Explain why the result is very nearly a straight line. Estimate ΔH^0 and ΔS^0 for the reaction. Compare these with the values obtained from Appendix 3 for 25°C.

23.   A and B are two gaseous optical isomers. Calculate the free energy change, ΔG, which occurs when one mole of A is converted to the equilibrium mixture of half a mole each of A and B at a temperature of 25°C. What happens to the free energy which is lost when this change occurs?

24.   According to the Clausius-Clapeyron Equation, the vapor pressure of a liquid varies with temperature as follows:

$$\log_e P = \frac{A}{T} + B$$

where A and B are constants. Show that this equation can be derived from equation (11.3) by assuming that ΔH^0 and ΔS^0 for the process

$$1 \text{ mole liquid} \rightarrow 1 \text{ mole gas}$$

are independent of temperature. Find A and B in terms of ΔH^0 and ΔS^0.

25.   Assume for the reaction

$$CO(g) + H_2O(g) \rightarrow CO_2(g) + H_2(g)$$

that ΔH^0 and ΔS^0 have the whole number values -10 kcal and -10 cal deg^{-1} respectively, irrespective of temperature. Calculate ΔG^0 and $\Delta G^0/T$ for this reaction at several temperatures in the range from 250°K to 5000°K. Use Table 11-2 to estimate the extent of this reaction when one mole each of CO and H_2O are allowed to react at one atmosphere total pressure at the various temperatures chosen. Finally, draw a graph showing the variation of the extent of the reaction with temperature.

26. Use the ΔG_f^0 tables in Appendix 4 to calculate ΔG^0 and K_p for the reaction

$$CO(g) + H_2O(g) \rightarrow CO_2(g) + H_2(g)$$

at $500°K$ and $1000°K$. Calculate the extent of the reaction at these two temperatures when one mole each of the reactants are mixed at a total pressure of one atmosphere. Compare these accurate values with the approximate values obtained in the previous problem.

PHYSICS

UNITS

The units employed in this text correspond to the usage of most U.S. textbooks in chemistry. To a large extent the c.g.s. system, based on the centimeter, the gram, and the second, is used. A major exception to this is the *calorie,* which is not a c.g.s. unit. The student should be made aware, though, that there is an internationally accepted system of units, usually called S.I. units, based on the meter, the kilogram, and the second, which has still to find favor with U.S. chemists. The unit of energy in this system is the *joule* rather than the calorie.

MECHANICS

Velocity The velocity of a body is the rate of change of its position with time. In the c.g.s. system it has units of centimeters per second or *cm sec*$^{-1}$

Acceleration The acceleration of a body is the rate of change of its velocity with time. In the c.g.s. system it has units of centimeters per second per second or *cm sec*$^{-2}$

Force The application of a force to a moving object accelerates it according to Newton's Law:

$$f = ma$$

where f is the applied force, m is the mass of the object, and a is the acceleration. A force of one *dyne* is defined as the force required to give one gram of matter an acceleration of one centimeter per second per second. A dyne thus has units of *g cm sec*$^{-2}$.

Energy and Work When a force moves through a distance, *work* is said to be done, or *energy* to be expended. When one dyne of force moves through one centimeter, one *erg* of work is said to be done. An erg thus has units of dyne-centimeters or *g cm*^{2} *sec*$^{-2}$. The erg is an inconveniently small unit for many purposes. The *joule,* equal to one kilogram-meter per second per second or 10^7

265

ergs, is a much larger unit.

Kinetic Energy Since the application of a force through a distance is required to give a body motion, the body must have energy because of this motion. This energy is called its *kinetic energy*. It can easily be proved that the kinetic energy, T, of a particle is given by the formula

$$T = \frac{1}{2}mv^2$$

where m is the mass of the particle and v is its velocity. Notice that the units in this equation are correct. If the mass is expressed in grams and the velocity in centimeters per second, the kinetic energy is given in the units $g\ cm^2\ sec^{-2}$; i.e., it is given in *ergs*.

HEAT FLOW

The Calorie When a body increases its temperature, it *gains* energy. Conversely, when it decreases its temperature, it *loses* energy. When a hot body and a cold body are allowed to interact, the one cools down and the other warms up until both are at the same temperature. Energy thus passes from one body to the other. Such energy is commonly called *heat energy* or just *heat*. The quantity of heat energy transferred depends on (a) the temperature difference between the two bodies, (b) their individual masses, and (c) a characteristic thermal property of each body called its *specific heat*, which we will define below.

In this book, heat energy is always measured in calories or kilocalories. A *calorie* is defined as the quantity of heat energy required to raise the temperature of 1 gram of water from $14.5°C$ to $15.5°C$. For most purposes we can regard the calorie as the amount of heat required to heat 1 gram of water through one degree Centigrade, whether from $1°C$ to $2°C$, or from $91°C$ to $92°C$, since the energy required is almost independent of temperature. A kilocalorie is equal to one thousand calories.

The calorie is, of course, related to other units of energy by means of a conversion factor. The conversion factor to ergs and joules is 1 calorie = 4.184 joule = 4.184×10^7 erg. Other conversion factors are given inside the front cover. It is quite often useful to remember that a calorie is about four times as big as a joule.

Specific Heats Although it requires one calorie of energy to raise the temperature of 1 gram of water $1°C$, it usually requires less energy if a gram of a substance other than water is used. Such a difference in behavior is measured by the *specific heat* of the substance.

The specific heat of a substance is the amount of energy required to raise the temperature of one gram of that substance by one degree Centigrade. For example, the specific heat of the element copper is 0.092 cal deg^{-1} g^{-1}. This means that it requires 0.092 calorie to raise the temperature of one gram of copper by one degree Centigrade. In general, the amount of heat required to raise the temperature of m grams of a substance is given by the relationship

$$Q = mC\Delta T \qquad \text{(A1.1)}$$

where C is the specific heat of the substance. For example, the number of calories required to heat up 15 grams of copper from 25°C to 35°C is given by

$$Q = 15 \text{ g} \times 0.092 \text{ cal deg}^{-1} \text{ g}^{-1} \times 10 \text{ deg}$$

$$= 13.8 \text{ cal}$$

In rough calculations we always assume that the specific heat of a substance does not vary with temperature, although in actual fact this is often not the case. Careful measurements on water, for instance, reveal that the specific heat is not exactly 1.0000 cal deg^{-1} g^{-1} at all temperatures. It has a value of 1.007 cal deg^{-1} g^{-1} at 0°C, decreases to a value of 0.998 cal deg^{-1} g^{-1} at 35°C, and then increases again to a value of 1.007 cal deg^{-1} g^{-1} at 100°C. The amount of heat required to heat a gram of water from 0°C to 100°C at one atmosphere pressure is actually 100.077 cal, whereas that predicted by equation (A1.1), assuming that C = 1.0000 cal deg^{-1} g^{-1}, is

$$Q = 1 \text{ g} \times 1.0000 \text{ cal deg}^{-1} \text{ g}^{-1} \times 100 \text{ deg}$$

$$= 100.000 \text{ cal}$$

For most purposes such a difference is negligible.

Molar Heat Capacities Chemists often find it useful to use a specific heat based on the mole rather than on the gram. The *molar heat capacity* of a substance is the quantity of energy required to raise the temperature of one mole of the substance by one degree Centigrade. The molar heat capacity of liquid water, for example, is the energy required to raise the temperature of 18.02 g of water through 1°C. It is thus 18.02 cal deg^{-1} mole^{-1} (at 15°C).

Molar heat capacities are often simply related to the gas constant R (R = 1.987 cal deg^{-1} mole^{-1}). They have the same units as R. We find, for instance, that many solid elements have a molar heat capacity which is close to the quantity 3R. Again, the molar heat capacity of any monatomic gas is 3R/2 at constant volume.

Appendix 2

THE EQUILIBRIUM CONSTANT

In a general gaseous reaction of the form

$$aA(g) + bB(g) + \ldots \rightleftharpoons mM(g) + nN(g) + \ldots \qquad \text{(A2.1)}$$

in which a moles of substance A, b moles of substance B, etc. react to form m moles of substance M, n moles of N, etc., we can write the equilibrium constant K_c as

$$K_c = \frac{[M]^m \cdot [N]^n \cdot \ldots}{[A]^a \cdot [B]^b \cdot \ldots} \qquad \text{(A2.2)}$$

where [A], [B], [M], [N], etc. are the equilibrium concentrations of the appropriate substances expressed in moles per liter.

As an example of such an equilibrium constant let us consider the reaction

$$2\,HI(g) \rightleftharpoons H_2(g) + I_2(g)$$

This equilibrium has been carefully studied at various temperatures. If HI gas is heated to $698.6°K$, it decomposes partly into H_2 and I_2. In one such experiment the following equilibrium concentrations were measured and found to be

$$[H_2] = 0.4953 \times 10^{-3} \text{ mole liter}^{-1}$$

$$[I_2] = 0.4953 \times 10^{-3} \text{ mole liter}^{-1}$$

$$[HI] = 3.655 \ \times 10^{-3} \text{ mole liter}^{-1}$$

The equilibrium constant for this reaction is given by

$$K_c = \frac{[H_2]\,[I_2]}{[HI]^2} \qquad \text{(A2.3)}$$

Substituting in the experimental values we have

$$K_c = \frac{(0.4953 \times 10^{-3})\,(0.4953 \times 10^{-3})}{(3.655 \times 10^{-3})^2}$$

$$= \frac{0.2453 \times 10^{-6}}{13.36 \times 10^{-6}}$$

$$= 1.832 \times 10^{-2}$$

This same equilibrium can also be studied by heating up unequal numbers of moles of H_2 and I_2 and measuring the equilibrium concentrations of these two species as well as the HI which is formed. In one such experiment, also at $698.6°K$, the equilibrium concentrations were found to be

$$[H_2] = 1.8313 \times 10^{-3} \text{ mole liter}^{-1}$$

$$[I_2] = 3.1292 \times 10^{-3} \text{ mole liter}^{-1}$$

$$[HI] = 17.671 \times 10^{-3} \text{ mole liter}^{-1}$$

Substituting these values into the expression for the equilibrium constant we then have

$$K_c = \frac{(1.8313 \times 10^{-3})\,(3.1292 \times 10^{-3})}{(17.671 \times 10^{-3})^2}$$

$$= \frac{5.729 \times 10^{-6}}{312.2 \times 10^{-6}}$$

$$= 1.835 \times 10^{-2}$$

in very good agreement with the value previously obtained. We thus expect that whatever the initial number of moles of H_2, I_2, and HI we mix together, and irrespective of the final volume or pressure, at a temperature of $698.6°K$ the final equilibrium concentrations will be interconnected by the relationship

$$\frac{[H_2]\,[I_2]}{[HI]^2} = 1.834 \times 10^{-2}$$

within the limits of experimental error.

In general, the equilibrium constant for a reaction varies with the temperature. For the equilibrium just considered, for instance, K_c increases with temperature. It is 2.05×10^{-2} at $731°K$ and 2.20×10^{-2} at $764°K$.

It is often convenient to use another equilibrium constant, which is expressed in terms of the *partial pressures* of the reactants and products rather than in terms of their concentrations. Such an equilibrium constant is usually represented by the symbol K_p. It is sometimes called the *pressure equilibrium constant*. For reaction (A2.1), K_p is defined by the expression

$$K_p = \frac{(p_M)^m \ (p_N)^n \cdots}{(p_A)^a \ (p_B)^b \ \ldots} \tag{A2.4}$$

where p_M, p_N, etc. are the partial pressures of the species M, N, etc. K_p is thus of the same form as K_c, with partial pressures now replacing concentrations.

Since, as we shall see, partial pressures and concentrations are related, there is also a relationship between K_p and K_c which may be derived in the following way.

If the gas mixture is an ideal one, its constituents will obey Dalton's Law of partial pressures. We can thus write for constituent A that

$$p_A V = n_A RT \tag{A2.5}$$

where n_A is the number of moles of A present, and V is the total volume of the gas mixture. Rewriting equation (A2.5), we obtain

$$p_A = RT\left(\frac{n_A}{V}\right)$$

But $n_A/V = [A]$, the number of moles of A per unit volume. Thus

$$p_A = RT \ [A] \tag{A2.6}$$

By similar arguments we also find that

$$p_B = RT \ [B], \, p_M = RT \ [M], \, p_N = RT \ [N], \ldots \tag{A2.7}$$

On substituting equations (A2.6) into (A2.4), the expression for K_p, we then have

$$K_p = \frac{(RT)^m \ (RT)^n \cdots}{(RT)^a \ (RT)^b \ldots} \times \frac{[M]^m \ [N]^n \cdots}{[A]^a \ [B]^b \ldots}$$

which shows us that K_p and K_c are related by:

$$K_p = K_c \ (RT)^{\Delta n}$$

where $\Delta n = (m + n + \ldots) - (a + b + \ldots)$. That is, Δn is equal to the change in the number of moles in reaction (A2.1). Note that Δn will be *positive* if the number of moles increases as the reaction proceeds from the left to the right. For example, in the reaction just considered,

$$2\,HI(g) \rightarrow H_2(g) + I_2(g)$$

the number of moles of reactant and products is the same, namely 2. Thus Δn is zero and $K_p = K_c$.

By contrast, for the reaction

$$N_2(g) + 3\,H_2(g) \rightarrow 2\,NH_3(g)$$

the number of moles of reactants is four while that of products is two. Thus $\Delta n = 2 - 4 = -2$. Accordingly,

$$K_p = \frac{K_c}{(RT)^2}$$

in this case.

Appendix 3

TABLES OF THERMODYNAMIC DATA AT 25°C

ΔH_f^0 = the standard enthalpy of formation of a substance from its elements at 25°C and 1 atm pressure in kcal mole^{-1}.

ΔG_f^0 = the standard free energy of formation of a substance from its elements at 25°C and 1 atm pressure in kcal mole^{-1}.

S^0 = the standard entropy of a substance at 25°C and 1 atm pressure in cal deg^{-1} mole^{-1}.

s = solid   l = liquid   g = gas

Substance	State	ΔH_f^0 (kcal mole^{-1})	ΔG_f^0 (kcal mole^{-1})	S^0 (cal deg^{-1} mole^{-1})
Aluminum				
Al	s	0.0	0.0	6.8
Al_2O_3	s	-400.5	-378.2	12.2
AlF_3	s	-359.5	-340.6	15.9
$AlCl_3$	s	-168.3	-150.3	26.5
$AlBr_3$	s	-126.0	--	--
$Al_2(SO_4)_3$	s	-822.4	-741.0	57.2
Antimony				
Sb	s	0.0	0.0	10.9
Sb_2O_5	s	-232.3	-198.2	29.9
$SbCl_3$	g	-75.0	-72.0	80.7
	s	-91.3	-77.4	44.0

Substance	State	ΔH_f^0	ΔG_f^0	S^0
Argon				
Ar	g	0.0	0.0	37.0
Arsenic				
As	s (α grey)	0.0	0.0	8.4
As_4O_6	s	-314.0	-275.5	51.2
As_2O_5	s	-221.0	-187.0	25.2
AsH_3	g	15.9	16.5	53.2
As_2S_3	s	-40.4	-40.3	39.1
Barium				
Ba	s	0.0	0.0	16.0
BaO	s	-133.4	-126.3	16.8
BaF_2	s	-286.9	-274.5	23.0
$BaCl_2$	s	-205.6	-193.8	30.0
$BaCl_2 2H_2O$	s	-349.3	-309.7	48.5
$BaSO_4$	s	-350.2	-323.4	31.6
$BaCO_3$	s	-291.3	-272.2	26.8
Beryllium				
Be	s	0.0	0.0	2.3
BeO	s	-146.0	-139.0	3.4
Bismuth				
Bi	s	0.0	0.0	13.6
Bi_2O_3	s	-137.2	-118.0	36.2
$BiCl_3$	g	-63.5	-61.2	85.7
	s	-90.6	-75.3	42.3
BiOCl	s	-87.7	-77.0	28.8
Boron				
B	s	0.0	0.0	1.4
B_2O_3	s	-304.2	-285.3	12.9
B_2H_6	g	8.5	20.7	55.5
H_3BO_3	s	-261.6	-231.6	21.2
BF_3	g	-271.8	-267.8	60.7
BCl_3	g	-96.5	-92.9	69.3
	l	-102.1	-92.6	49.3
BBr_3	g	-49.2	-55.6	77.5
	l	-57.3	-57.0	54.9
Bromine				
Br_2	l	0.0	0.0	36.4
	g	7.4	0.7	58.6
Br	g	26.7	19.7	41.8
BrCl	g	3.50	-0.23	57.4
Cadmium				
Cd	s	0.0	0.0	12.4
CdO	s	-61.7	-54.6	13.1
$CdCl_2$	s	-93.6	-82.2	27.6
CdS	s	-38.7	-37.4	15.5
$CdSO_4$	s	-223.1	-196.7	29.4
Calcium				
Ca	s	0.0	0.0	9.9
CaO	s	-151.9	-144.4	9.5
$Ca(OH)_2$	s	-235.8	-214.3	18.2

Substance	State	ΔH_f^0	ΔG_f^0	S^0
$CaCl_2$	s	−190.0	−179.3	27.2
$CaSO_4$	s	−342.4	−315.6	25.5
$CaSO_4 \cdot 2H_2O$	s	−483.1	−429.2	46.4
CaC_2	s	−15.0	−16.2	16.8
$CaCO_3$	s	−288.4	−269.8	22.2
Carbon				
C	s (graphite)	0.0	0.0	1.37
	s (diamond)	0.45	0.69	0.57
	g	171.3	160.4	37.8
CO	g	−26.4	−32.8	47.2
CO_2	g	−94.1	−94.3	51.1
CH_4	g	−17.9	−12.1	44.5
C_2H_6	g	−20.2	−7.9	54.8
C_2H_4	g	12.5	16.3	52.5
C_2H_2	g	54.2	50.0	48.0
C_3H_8	g	−24.8	−5.6	64.5
$n\text{-}C_4H_{10}$	g	−29.8	−3.7	74.1
$iso\text{-}C_4H_{10}$	g	−31.5	−4.3	70.4
$n\text{-}C_5H_{12}$	g	−35.0	−2.0	83.3
C_6H_6	g	19.8	31.0	64.3
$n\text{-}C_6H_{14}$	g	−40.0	−0.1	92.8
$n\text{-}C_7H_{16}$	g	−44.9	1.9	102.3
HCOOH	l	−101.5	−86.4	30.8
CH_3OH	g	−48.0	−38.7	57.3
	l	−57.0	−39.8	30.3
CF_4	g	−221.0	−210.0	62.5
CCl_4	g	−24.6	−14.5	74.0
	l	−32.4	−15.6	51.7
$COCl_2$	g	−52.3	−48.9	67.7
CH_3Cl	g	−19.3	−13.7	56.0
	l	−24.3	−12.3	34.6
CH_2Cl_2	g	−22.1	−15.8	64.6
	l	−29.0	−16.1	42.5
$CHCl_3$	g	−24.6	−16.8	70.6
	l	−32.1	−17.6	48.2
CH_3I	g	3.1	3.5	60.7
	l	−3.7	3.2	39.0
CS_2	g	28.0	16.0	56.8
	l	21.4	15.6	36.2
Cesium				
Cs	s	0.0	0.0	19.8
Cs_2O	s	−75.9	−−	−−
CsF	s	−126.9	−119.7	19.8
CsCl	s	−103.5	−96.8	23.9
CsBr	s	−94.3	−91.6	29.0
CsI	s	−80.5	−79.7	31.0
Chlorine				
Cl_2	g	0.0	0.0	53.3
Cl	g	29.1	25.3	39.5
Cl_2O	g	19.2	23.4	63.6
ClF	g	−11.9	−12.3	52.1

Substance	State	ΔH_f^0	ΔG_f^0	S^0
ClF_3	g	−38.0	−28.4	67.3
Chromium				
Cr	s	0.0	0.0	5.7
Cr_2O_3	s	−272.4	−252.9	19.4
CrO_3	s	−140.9	−−	−−
$CrCl_2$	s	−94.5	−85.1	27.6
$CrCl_3$	s	−133.0	−116.2	29.4
Cobalt				
Co	s	0.0	0.0	7.2
CoO	s	−56.9	−51.2	12.7
$CoCl_2$	s	−74.7	−64.5	26.1
Copper				
Cu	s	0.0	0.0	7.9
Cu_2O	s	−40.3	−34.9	22.3
CuO	s	−37.6	−31.0	10.2
CuCl	s	−32.8	−28.7	20.6
$CuCl_2$	s	−52.6	−42.0	25.8
Cu_2S	s	−19.0	−20.6	28.9
CuS	s	−12.7	−12.8	15.9
$CuSO_4$	s	−184.4	−158.2	26.0
$CuSO_4 \cdot 5H_2O$	s	−544.9	−449.3	71.8
Fluorine				
F_2	g	0.0	0.0	48.4
F	g	18.9	14.7	37.9
F_2O	g	−5.2	−1.1	59.1
Germanium				
Ge	s	0.0	0.0	7.4
$GeCl_4$	l	−127.1	−110.6	58.7
Gold				
Au	s	0.0	0.0	11.3
$AuCl_3$	s	−28.1	−−	−−
Helium				
He	g	0.0	0.0	30.1
Hydrogen				
H_2	g	0.0	0.0	31.2
H	g	52.1	48.6	27.4
H_2O	g	−57.8	−54.6	45.1
	l	−68.3	−56.7	16.7
H_2O_2	g	−32.6	−25.3	55.6
	l	−44.9	−28.8	26.2
HF	g	−64.8	−65.3	41.5
HCl	g	−22.1	−22.8	44.6
HBr	g	−8.7	−12.8	47.5
HI	g	6.3	0.5	49.4
H_2S	g	−4.9	−8.0	49.2
H_2Se	g	7.1	3.8	52.3
Iodine				
I_2	s	0.0	0.0	27.8
	g	14.9	4.6	62.3
I	g	25.5	16.8	43.2
ICl	g	4.2	−1.3	59.1
	l	−5.7	−3.3	32.3

Substance	State	ΔH_f^0	ΔG_f^0	S^0
ICl_3	s	-21.4	-5.3	40.0
IBr	g	9.8	0.9	61.8
Iron				
Fe	s	0.0	0.0	6.5
Fe_2O_3	s	-197.0	-177.4	20.9
$FeCl_2$	s	-81.7	-72.3	28.2
$FeCl_3$	s	-95.5	-79.8	34.0
FeS_2	s	-42.5	-39.8	12.7
Krypton				
Kr	g	0.0	0.0	39.2
Lead				
Pb	s	0.0	0.0	15.5
PbO	s (red)	-52.3	-45.2	15.9
	s (yellow)	-51.9	-44.9	16.4
PbO_2	s	-66.3	-51.9	16.4
Pb_3O_4	s	-171.7	-143.7	50.5
PbF_2	s	-158.7	-147.5	26.4
$PbCl_2$	s	-85.9	-75.1	32.5
$PbBr_2$	s	-66.6	-62.6	38.6
PbI_2	s	-41.9	-41.5	41.8
PbS	s	-24.0	-23.6	21.8
$PbSO_4$	s	-219.9	-194.4	35.5
Lithium				
Li	s	0.0	0.0	6.7
	g	37.1	29.2	33.1
Li_2O	s	-142.4	-133.8	9.1
LiF	s	-146.3	-139.6	8.6
LiCl	s	-97.7	-91.9	13.9
LiBr	s	-83.7	-81.2	16.5
Li_2SO_4	s	-342.8	-316.0	29.0
Li_2CO_3	s	-290.5	-270.7	21.6
Magnesium				
Mg	s	0.0	0.0	7.8
MgO	s	-143.8	-136.1	6.4
$Mg(OH)_2$	s	-221.0	-199.3	15.1
MgF_2	s	-263.5	-250.8	13.7
$MgCl_2 \cdot 6H_2O$	s	-597.4	505.7	87.5
$MgSO_4 \cdot 7H_2O$	s	-808.7	$--$	$--$
$Mg(NO_3)_2$	s	-188.7	-140.6	39.2
$MgCO_3$	s	-266.0	-246.0	15.7
Manganese				
Mn	s	0.0	0.0	7.6
MnO	s	-92.0	-86.8	14.4
MnO_2	s	-124.3	-111.2	12.7
$MnCl_2$	s	-115.0	-105.3	28.3
$MnBr_2$	s	-92.0	$--$	$--$
$MnSO_4 \cdot 4H_2O$	s	-539.7	$--$	$--$
Mercury				
Hg	l	0.0	0.0	18.2
	g	14.6	7.6	41.8

Substance	State	ΔH_f^0	ΔG_f^0	S^0
HgO	s (red)	−21.71	−13.99	16.8
	s (yellow)	−21.62	−13.96	17.0
Hg_2Cl_2	s	−63.4	−50.4	46.0
$HgCl_2$	s	−53.6	−42.7	34.9
HgS	s (red)	−13.9	−12.1	19.7
	s (black)	−12.8	−11.4	21.1
Molybdenum				
Mo	s	0.0	0.0	6.8
MoO_3	s	−178.1	−159.7	18.7
Neon				
Ne	g	0.0	0.0	34.9
Nickel				
Ni	s	0.0	0.0	7.1
NiO	s	−57.3	−50.6	9.1
$NiCl_2 \cdot 6H_2O$	s	−502.7	−409.5	82.3
$Ni(CO)_4$	g	−144.1	−140.4	98.1
Nitrogen				
N_2	g	0.0	0.0	45.8
N	g	113.0	108.9	36.6
NO	g	21.6	20.7	50.3
NO_2	g	7.9	12.3	57.4
N_2O	g	19.6	24.9	52.5
N_2O_4	g	2.2	23.4	72.7
N_2O_5	s	−10.3	27.2	42.6
NH_3	g	−11.0	−3.9	46.0
N_2H_4	g	22.8	38.1	57.0
	l	12.1	35.7	29.0
HNO_3	l	−41.6	−19.3	37.2
NH_4NO_3	s	−87.4	−44.0	36.1
NF_2	g	10.3	13.9	59.7
NF_3	g	−29.8	−20.0	62.3
N_2F_4	g	−1.7	19.4	72.0
NOF	g	−15.9	−12.2	59.3
NOCl	g	12.4	15.8	62.5
NH_4Cl	s	−75.2	−48.5	22.6
NOBr	g	19.6	19.7	65.4
$(NH_4)_2SO_4$	g	−282.2	−215.6	52.6
Oxygen				
O_2	g	0.0	0.0	49.0
O	g	59.6	55.4	38.5
O_3	g	34.1	39.0	57.1
Phosphorus				
P	s (white)	0.0	0.0	9.8
	s (red)	−4.2	−2.9	5.4
	g	75.2	66.5	39.0
P_4O_{10}	s	−713.2	−644.8	54.7
PH_3	g	1.3	3.2	50.2
H_3PO_4	s	−305.7	−267.5	26.4
PCl_3	g	−68.6	−64.0	74.5
	l	−76.4	−65.1	51.9
PCl_5	g	−89.6	−72.9	87.1

Substance	State	ΔH_f^0	ΔG_f^0	S^0
PCl_5	s	−106.0		
$POCl_3$	g	−133.5	−122.6	77.8
	l	−142.7	−124.5	53.2
PBr_3	g	−33.3	−38.9	83.2
	l	−44.1	−42.0	57.4
Potassium				
K	s	0.0	0.0	15.2
K_2O	s	−86.4	−77.0	23.5
KF	s	−134.5	−127.4	15.9
KCl	s	−104.2	−97.6	19.8
$KClO_3$	s	−93.5	−69.3	34.2
KBr	s	−93.7	−90.6	23.0
$KBrO_3$	s	−79.4	−58.2	35.6
KI	s	−78.3	−77.0	24.9
KIO_3	s	−121.5	−101.7	36.2
K_2SO_4	s	−342.7	−314.6	42.0
KNO_3	s	−117.8	−94.0	31.8
$KMnO_4$	s	−194.4	−170.6	41.0
KOH	s	−101.8	−−	−−
Radon				
Rn	g	0.0	0.0	42.1
Rubidium				
Rb	s	0.0	0.0	16.6
RbF	s	−131.3	−124.0	18.0
RbCl	s	−102.9	−96.0	21.9
RbBr	s	−93.0	−90.4	25.9
RbI	s	−78.5	−77.8	28.2
Selenium				
Se	s (black)	0.0	0.0	10.1
SeF_6	g	−267.0	−243.0	75.0
SeO_2	s	−53.9	−−	−−
Silicon				
Si	s	0.0	0.0	4.5
SiO_2	s (quartz)	−217.7	−204.8	10.0
SiH_4	g	8.2	13.6	48.9
SiF_4	g	−386.0	−375.9	67.5
$SiCl_4$	g	−157.0	−147.5	79.0
	l	−164.2	−148.2	57.3
Silver				
Ag	s	0.0	0.0	10.2
Ag_2O	s	−7.4	−2.7	29.0
AgF	s	−48.9	−−	−−
AgCl	s	−30.4	−26.2	23.0
AgBr	s	−24.0	−23.2	25.6
AgI	s	−14.8	−15.8	27.6
Ag_2S	s	−7.8	−9.7	34.4
Ag_2SO_4	s	−171.1	−147.8	47.9
$AgNO_3$	s	−29.7	−8.0	33.7

Substance	State	ΔH_f^0	ΔG_f^0	S^0
Sodium				
Na	s	0.0	0.0	12.2
Na_2O	s	−99.4	−90.0	17.4
NaOH	s	−102.0	−91.0	15.3
NaF	s	−136.5	−129.3	12.3
NaCl	s	−98.6	−92.2	17.4
NaBr	s	−86.0	−82.9	20.8
NaI	s	−68.8	−67.5	23.5
$Na_2SO_4 \cdot 10H_2O$	s	−1033.5	−870.9	141.7
$NaNO_3$	s	−111.5	−87.5	27.8
Na_2CO_3	s	−270.3	−250.4	32.5
$Na_2CO_3 \cdot 10H_2O$	s	−975.6	−−	−−
$NaHCO_3$	s	−226.5	−203.6	24.4
Strontium				
Sr	s	0.0	0.0	13.0
SrO	s	−141.1	−133.8	13.0
$SrCl_2$	s	−198.0	−186.7	28.0
$SrSO_4$	s	−345.3	−318.9	29.1
$SrCO_3$	s	−291.2	−271.9	23.2
Sulfur				
S	s (rhombic)	0.0	0.0	7.6
	s (monoclinic)	0.08	−−	−−
	g	30.7	18.9	54.5
S_2	g	66.6	56.9	40.1
S_8	g	24.4	11.9	103.0
SO_2	g	−70.9	−71.7	59.3
SO_3	s	−108.6	−88.2	12.5
	g	−94.6	−88.7	61.3
H_2SO_4	l	−194.5	−164.9	37.5
SF_6	g	−289.0	−264.2	69.7
Tellurium				
Te	s	0.0	0.0	11.9
TeO_2	s	−77.1	−64.6	19.0
TeF_6	g	−315.0	−292.0	80.7
Tin				
Sn	s (white)	0.0	0.0	12.3
	s (grey)	−0.5	0.0	10.5
SnO	s	−68.3	−61.4	13.5
SnO_2	s	−138.8	−124.2	12.5
$SnCl_4$	l	−122.2	−105.2	61.8
$SnCl_2 \cdot 2H_2O$	s	−220.2	−−	−−
Titanium				
Ti	s	0.0	0.0	7.3
TiO_2	s	−225.8	−212.6	12.0
$TiCl_4$	l	−192.2	−176.2	60.3
Tungsten				
W	s	0.0	0.0	7.8
WO_3	s	−201.5	−182.6	18.1
Uranium				
U	s	0.0	0.0	12.0
UO_2	s	−270.0	−257.0	18.6

Substance	State	ΔH_f^0	ΔG_f^0	S^0
UF_6	g	−505.0	−485.0	90.8
Vanadium				
V	s	0.0	0.0	7.0
V_2O_5	s	−370.6	−339.3	31.3
Xenon				
Xe	g	0.0	0.0	40.5
XeF_2	g	−25.9	−−	−−
XeF_4	g	−51.5	−−	−−
	s	−62.5	−−	−−
XeF_6	g	−71.3	−−	−−
Zinc				
Zn	s	0.0	0.0	10.0
ZnO	s	−83.2	−76.1	10.4
$ZnCl_2$	s	−99.2	−88.3	26.6
$ZnBr_2$	s	−78.6	−74.6	33.1
ZnI_2	s	−49.7	−49.9	38.5
ZnS	s (zinc blende)	−49.2	−48.1	13.8
$ZnSO_4$	s	−234.9	−209.0	28.6
$ZnSO_4 \cdot 7H_2O$	s	−735.6	−612.6	92.9
$ZnCO_3$	s	−194.3	−174.9	19.7
Zirconium				
Zr	s	0.0	0.0	9.3
ZrO_2	s	−263.0	−249.2	12.0
$ZrCl_4$	s	−234.4	−212.7	43.4

Most thermodynamic data recorded above have been taken from the National Bureau of Standards Circular 500 as updated by the N.B.S. Technical Notes 270-3, 270-4, and 270-5.

STANDARD MOLAR FREE ENERGIES OF FORMATION FROM 300°K TO 1000°K

Sources: 1) *JANAF Thermochemical Data,* Dow Chemical Co.
2) *The Chemical Thermodynamics of Organic Compounds:* Stall, Westrum, and Sinke, John Wiley and Sons, Inc.

See tables on following pages.

Elements and Inorganic Compounds

	300°K	400°K	500°K	600°K	700°K	800°K	900°K	1000°K
Al(s)	0.00	0.00	0.00	0.00	0.00	0.00	0.00	0.19
Al_2O_3(s)	-377.94	-370.42	-362.89	-355.39	-347.92	-340.48	-333.07	-325.30
Br_2(g)	0.70	0.00	0.00	0.00	0.00	0.00	0.00	0.00
C(graphite)	0.00	0.00	0.00	0.00	0.00	0.00	0.00	0.00
CO(g)	-32.82	-34.98	-37.14	-39.31	-41.47	-43.61	-45.74	-47.86
CO_2(g)	-94.27	-94.34	-94.40	-94.46	-94.51	-94.56	-94.60	-94.63
$COCl_2$(g)	-49.40	-48.27	-47.15	-46.05	-44.93	-43.83	-42.73	-41.64
Cl_2(g)	0.00	0.00	0.00	0.00	0.00	0.00	0.00	0.00
F_2(g)	0.00	0.00	0.00	0.00	0.00	0.00	0.00	0.00
Fe(s)	0.00	0.00	0.00	0.00	0.00	0.00	0.00	0.00
Fe_2O_3(s)	-177.60	-171.08	-164.68	-158.41	-152.25	-146.21	-140.25	-134.36
H_2(g)	0.00	0.00	0.00	0.00	0.00	0.00	0.00	0.00
HCl(g)	-22.78	-23.01	-23.22	-23.42	-23.60	-23.77	-23.94	-24.09
HI(g)	0.34	-1.53	-2.41	-2.62	-2.81	-3.00	-3.18	-3.35
H_2O(g)	-54.62	-53.52	-52.36	-51.16	-49.92	-48.65	-47.35	-46.04
I_2(g)	4.57	1.32	0.00	0.00	0.00	0.00	0.00	0.00
Mg(s)	0.00	0.00	0.00	0.00	0.00	0.00	0.00	0.18
MgO(s)	-135.94	-133.35	-130.76	-128.19	-125.62	-123.06	-120.51	-117.78
N_2(g)	0.00	0.00	0.00	0.00	0.00	0.00	0.00	0.00
NH_3(g)	-3.92	-1.47	1.11	3.78	6.51	9.29	12.10	14.93
NO(g)	20.69	20.39	20.10	19.80	19.49	19.19	18.89	18.59
NO_2(g)	12.27	13.75	15.26	16.78	18.30	19.83	21.36	22.88
N_2O_4(g)	23.49	30.62	37.76	44.88	51.96	58.99	65.99	72.94
NOBr(g)	19.62	20.40	21.55	22.68	23.82	24.95	26.08	27.20
NOCl(g)	16.07	17.22	18.38	19.53	20.67	21.81	22.94	24.06
Na(l)	0.12	0.00	0.00	0.00	0.00	0.00	0.00	0.00
Na(s)	0.00	0.05	0.21	0.36	0.49	0.61	0.72	0.81
NaBr(s)	-83.46	-81.69	-79.48	-77.29	-75.14	-73.01	-70.92	-68.86
NaCl(s)	-91.75	-89.55	-87.25	-84.97	-82.73	-80.52	-78.34	-76.20
Na_2O(s)	-90.55	-87.33	-83.86	-80.41	-77.00	-73.64	-70.33	-67.06
O_2(g)	0.00	0.00	0.00	0.00	0.00	0.00	0.00	0.00
PCl_3(g)	-57.74	-56.65	-55.55	-54.46	-53.36	-57.57	-54.58	-51.61
PCl_5(g)	-66.42	-61.29	-56.20	-51.16	-46.16	-46.50	-39.68	-32.91
SO_2(g)	-71.75	-71.95	-71.92	-71.79	-71.56	-72.57	-70.82	-69.07
SO_3(g)	-88.65	-86.60	-84.30	-81.92	-79.44	-78.21	-74.23	-70.26

Gaseous Organic Compounds

	$300°K$	$400°K$	$500°K$	$600°K$	$700°K$	$800°K$	$900°K$	$1000°K$
CH_4, methane	-12.11	-10.07	-7.85	-5.51	-3.06	-0.56	1.99	4.58
$HC{\equiv}CH$, acetylene	49.97	48.57	47.19	45.83	44.50	43.18	41.88	40.61
$H_2C{=}CH_2$, ethylene	16.31	17.69	19.25	20.92	22.68	24.49	26.35	28.25
$H_3C{-}CH_3$, ethane	-7.79	-3.45	1.16	5.96	10.90	15.91	21.00	26.13
$CH_3CH_2CH_3$, propane	-5.50	1.19	8.23	15.50	22.93	30.45	38.05	45.71
C_4H_{10}, n-butane	-3.94	5.10	14.55	24.27	34.19	44.21	54.33	64.50
C_4H_{10}, isobutane	-4.83	4.58	14.39	24.48	34.75	45.13	55.60	66.11
C_5H_{10}, 1-pentene	19.06	27.37	36.07	45.04	54.20	63.47	72.83	82.24
C_5H_{10}, cyclopentane	9.40	19.06	29.24	39.75	50.49	61.35	72.30	83.31
CH_3OH, methanol	-38.78	-35.54	-32.11	-28.52	-24.84	-21.10	-17.30	-13.46
CH_3Cl, methyl chloride	-15.00	-13.01	-10.88	-8.05	-6.34	-3.98	-1.59	0.83
CH_3Br, methyl bromide	-6.71	-5.09	-2.97	-0.75	1.53	3.86	6.23	8.63
CH_3I, methyl iodide	3.74	4.07	5.51	7.69	9.94	12.23	14.55	16.89

ANSWERS
TO
PROBLEMS

Chapter 1

1. $1/52$

2. $4/52 = 1/13$

3. $13/52 = 1/4$

4. $1/6$

5. $1/6 \times 1/6 = 1/36$

6. $2/36 = 1/18$

7. $4/36 = 1/9$

8. $(1/2)^6 = 1/64; 1/2$

9. $1/4 \times 2/4 = 1/8$

10. a) $K_c = 4$
    b) $K_c = 9$

11. a) 1 mole HOD, 1/2 mole D_2O, 1/2 mole H_2O
    b) 1/4 mole NH_3, 1/4 mole ND_3, 3/4 mole NH_2D, 3/4 mole NHD_2

12. a) 1/3 mole H_2, 1 1/3 mole D_2, 1 1/3 mole HD
    b) 1/9 mole NH_3, 8/9 mole ND_3, 2/3 mole NH_2D, 1 1/3 mole NHD_2

13. a) 2/3
    b) 1/3

Chapter 2

2.1.2 One distribution has two particles singly excited ($W = 6$). The other has one particle doubly excited ($W = 4$).

2.2.1 5040, 40320

2.2.2 Yes

2.2.3 $y > x$

2.2.4 8

2.2.5 n

2.2.6 $1/52! = 1.24 \times 10^{-68}$

2.3.1 $W = 42$, $W = 105$, $W = 35$. The most probable distribution has two particles singly excited and one doubly excited. Note that the populations form a geometric series.

2.3.2 The second most probable distribution is

Level	Population
3	0
2	3
1	5
0	7

2.4.1

Level	Populations		
3	1	–	–
2	–	1	–
1	–	1	3
0	$(n-1)$	$(n-2)$	$(n-3)$
Distribution	X	Y	Z

Z is $(n-2)/3$ times more probable than Y.

2.4.4 $\dfrac{W_A}{W_H} = \dfrac{n_1+1}{(n_2)(n_2-1)} \approx \dfrac{n_1}{(n_2)^2} = 94.$

2.5.1 At $100°K$; $n_1 = 5.64 \times 10^5$, $n_2 = 0.0$, $n_3 = 0.0$
At $200°K$; $n_1 = 5.83 \times 10^{14}$, $n_2 = 5.64 \times 10^5$, $n_3 = 0.0$
At $300°K$; $n_1 = 5.89 \times 10^{17}$, $n_2 = 5.76 \times 10^{11}$, $n_3 = 5.64 \times 10^5$

2.5.2 H_2, 1.93×10^{-9}; D_2, 5.41×10^{-7}; O_2, 5.47×10^{-4}; F_2, 1.37×10^{-2}; I_2, 3.57×10^{-1}.

2.5.3 Rotationally, $n_1/n_0 = 0.998$
Vibrationally, $n_1/n_0 = 0.068$

2.5.4 H_2, 0.564; D_2, 0.752; O_2, 0.981; Br_2, 0.9999.

2.5.6 $298°K = 25°C$

2.5.7 $860°K$

2.5.10 5.76×10^{22} particles

Chapter 3

1. a) 4.77×10^{-34} cm.
   b) 2.85×10^{-9} cm = 0.285 Å
   c) 3.64×10^{-8} cm = 3.64 Å

2. 7.27×10^9 cm sec^{-1}

3. $0.132, 0.528, 1.19, 2.11, 3.30, 4.76$ cal mole^{-1}

4. $0.297, 1.19, 2.68, 4.76, 7.43, 10.70$ cal mole^{-1}

5. 4.72×10^7 kcal mole^{-1}

6. a) $xe < Kr < Ne < He$
   b) $SF_6 < HI < F_2 < CH_4$

7. About 500 levels

8. 72.0 cal mole^{-1} (from equation (3.9)), 48.6 cal mole^{-1} (from equation (3.10))

9. 0.213 cal mole^{-1} (I_2), 350 cal mole^{-1} (H_2)

10. a) $Cl_2 < F_2 < O_2 < N_2$
    b) $CI_4 < CBr_4 < CF_4 < CH_4$
    c) $SF_6 < H_2S < H_2O < H_2$

12. $1/2 \times 8.25 = 4.1$ kcal mole^{-1}

13. F_2 has the more closely spaced levels

14. About 6

15. 72 kcal mole^{-1}

Chapter 4

1. $K_c(100°K) = 9.5 \times 10^{-2}$, $K_c(200°K) = 0.92$, $K_c(300°K) = 2.01$, $K_c(400°K) = 2.98$. Limiting value for K_c is 10.

2. $K_c(100°K) = 6 \times 10^{-3}$, $K_c(200°K) = 2.5 \times 10^{-2}$, $K_c(300°K) = 4.1 \times 10^{-2}$, $K_c(400°K) = 5.0 \times 10^{-2}$. Limiting value for K_c is 1×10^{-1}.

4. $K_c(298°K) = 1.41 \times 10^{-7}$, $K_c(500°K) = 1.28 \times 10^{-4}$, $K_c(1000°K) = 1.96 \times 10^{-2}$

5. a) $K_c = 2.53$
  b) $K_c = 0.55$
  c) $K_c = 1.4 \times 10^{-7}$
  d) $K_c = 1.6 \times 10^{-73}$

7. a) ii
  b) i
  c) v
  d) iii
  e) iv

8. Under no circumstances

9. Isobutane has the lower ground state.
  n-Butane has the more closely spaced levels.

10. m-Xylene has the lower ground state. Since K_c is less than one at both temperatures, not enough information is given to say which isomer has the more closely-spaced levels.

Chapter 5

1. Endothermic

2. Density, molar heat capacity, refractive index, dielectric constant, viscosity.

3. 0.289 kcal

4. 1.37 cal deg^{-1} mole^{-1} (100-150°K)
  1.64 cal deg^{-1} mole^{-1} (150-200°K)
  1.91 cal deg^{-1} mole^{-1} (200-250°K)

5. 117 kcal mole^{-1}

6. a) -2 kcal
  b) -29 kcal
  c) -194 kcal
  d) -137 kcal
  e) -53 kcal
  f) $+32$ kcal
  g) -53 kcal
  h) -141 kcal
  i) $+55$ kcal
  j) -10 kcal

7. -115 kcal (F_2); -25 kcal (Cl_2); -9 kcal (Br_2); $+7$ kcal (I_2)

8. a) v
  b) iv
  c) iii
  d) i
  e) ii

9. +170.3 kcal

10. In diamond each C atom is singly bonded to *four* other C atoms. When diamond sublimes, all these bonds must be broken. This means breaking *two* bonds per C atom, since each bond is attached to *two* atoms of carbon.

11. a) 24.45 liters
b) 24.45 liter atmospheres
c) 592.2 cal
d) $\Delta E = -37.0$ kcal

12. $w = 6.50 \times 10^{-3}$ cal
$q = 660$ cal
$\Delta E = 660$ cal

13. 72 kcal mole^{-1}
O_3 is a resonance hybrid:

Chapter 6

1. a) Kr
b) HCl
c) HI
d) CF_4
e) H_2S
f) NaF
g) $NaNO_3$

2. NH_3 has a larger number of ways of vibrating than has N_2. Some of these vibrations are quite "loose."

3. S_{298}^0 for CH_4 is 44.5 cal deg^{-1} mole^{-1} (not 74.5 cal deg^{-1} mole^{-1}). S_{298}^0 for I_2 is 62.3 cal deg^{-1} mole^{-1} (not 27.8 cal deg^{-1} mole^{-1}).

4. a) $H_2 < F_2 < Cl_2 < Br_2 < I_2$
b) $CH_3D < CH_3F < CH_3Cl < CH_3Br < CH_3I$
c) $H_2 < N_2 < F_2 < S_2$
d) $TiN < CaO < CaS < KCl < KI$
e) $He < Ne < HF < D_2O < ND_3$
f) $H_2O < H_2S < H_2O_2 < S_2Cl_2$

5. a) +4.7 cal deg^{-1}
b) -93.2 cal deg^{-1}
c) +5.8 cal deg^{-1}
d) +1.5 cal deg^{-1}
e) +44.4 cal deg^{-1}

6. a) negative   e) negative
    b) negative   f) small
    c) · positive   g) positive
    d) small   h) small

7. The S^0_{298} value for NCl_3 must lie between that for PCl_3 (74.5 cal deg^{-1} mole^{-1}) and NF_3 (62.3 cal deg^{-1} mole^{-1}) and is closer to the former value. 70 cal deg^{-1} would not be a bad guess.

8. $\Delta S^0 = 0$

9. W is always greater than 1, so k log$_e$W is always greater than zero.

10. $S = 24.0$ cal deg^{-1}

11. $W = 10^{(6.6 \times 10^{24})} = (10)^{(10^{24.82})}$

12. 19.8 cal deg^{-1}

13. 20.2 cal deg^{-1}

14. 34.9 cal deg^{-1}

15. S^0_{trans} $(O_2) = 36.3$ cal deg^{-1} ; S^0_{rot} $= 49.0 - 36.3 - 2.2$
                                      $= 10.5$ cal deg^{-1}

16. $S_{elec} = R \log_e 2 = 1.38$ cal deg^{-1}

17. $W = 2^N$; $S = R \log_e 2 = 1.38$ cal deg^{-1}

18. $\Delta Z = q/2.303$ kT

19. $S = R\left\{-(1/3) \log_e(1/3) - (2/3)\log_e(2/3)\right\} = 1.26$ cal deg^{-1}

20. i) $\Delta S = 0$ at $0°K$, $\Delta S = R \log_e 2$ at very high temperatures
   ii) $\Delta S = 0$ at $0°K$, $\Delta S = -R \log_e 2$ at very high temperatures
   iii) $\Delta S = 0$ at $0°K$, $\Delta S = R \log_e 2$ at very high temperatures
   iv) $\Delta S = 0$ at all temperatures
   v) $\Delta S = 0$ at all temperatures

21. The difference in entropy is too large to be explained by the very small difference in mass. Most of the difference is caused by a symmetry effect not explained in the text. CH_4 has more *rotational symmetry* than NH_3. There are 12 indistinguishable yet different orientations of CH_4, but only 3 such orientations for NH_3. The existence of this symmetry reduces the rotational contribution to the entropy by a factor of R log$_e$12 for CH_4 and by R log$_e$3 for NH_3. The effect depends on the Pauli principle and is too complicated to explain briefly.

Chapter 7

1. a) Forward reaction at low T, reverse reaction at high T

   b) Reverse reaction at all temperatures
   c) Forward reaction at all temperatures
   d) Reverse reaction at low T, forward reaction at high T
   e) Forward reaction at low T, reverse reaction at high T
   f) Forward reaction at low T, reverse reaction at high T
   g) Forward reaction at low T, uncertain at high T
   h) Reverse reaction at low T, forward reaction at high T
   i) Reverse reaction at low T, forward reaction at high T
   j) Forward reaction at all temperatures
   k) Forward reaction at all temperatures

2. Reactants have higher internal energy and higher standard entropy.

3. a) Reactants favored at low T, products at high T
   b) Reactants favored at all temperatures
   c) Products favored at all temperatures
   d) Products favored at low T, reactants at high T

4. Calcite at high temperatures, aragonite at low temperatures

5. Graphite is the stable form at all temperatures (at least at one atmosphere pressure).

6. i) Gray tin has the lower standard molar internal energy
   ii) Gray tin has the lower standard molar entropy

7. The vapor has both a higher molar energy and entropy.

8. Since a solid and a liquid are mixed to form a solid, ΔS for the setting process is presumably negative. The setting process can only occur if ΔE is negative, i.e., if it is exothermic.

9. Fluorine forms very strong bonds with most of the non-metals. As a result ΔE for the formation of most fluorides is negative.

10. ΔE is negative and ΔS is positive for the decomposition of this compound into its elements. It is thus unstable with respect to decomposition at all temperatures.

11. The reaction $2 \ HNF_2 \rightarrow N_2 + 2HF + F_2$ is exothermic and results in species of higher entropy.

12. ΔS for the reaction $2 \ XeO_3 \rightarrow 2 \ Xe + 3 \ O_2$ is positive and ΔE is negative (-231 kcal).

Chapter 8

1. $26.4°C$

2. a) 25 liter-atmospheres
   b) 605.5 cal

3. a) -0.59 kcal
   b) $+0.59$ kcal
   c) $+0.70$ kcal
   d) very small
   e) $+1.987$ kcal

4. $+10.6$ kcal

5. -1228 kcal

6. $\Delta H = +5.1$ kcal (Interestingly, this is an *endothermic* acid-base reaction.)

7. 18.89 kcal deg^{-1}

8. $\Delta E = -170.6$ kcal; $\Delta H = -170.3$ kcal

9. $\Delta H_f^0 = +11.5$ kcal

10. *Yellow* phosphorous is taken as zero.

11. a) -70.9 kcal
   b) -22.7 kcal
   c) $+14.5$ kcal
   d) $+6.6$ kcal
   e) -143.8 kcal
   f) $+21.0$ kcal
   g) -204.0 kcal
   h) $+12.6$ kcal
   i) -136.6 kcal
   j) -284.8 kcal

12. a) -2.3 kcal
   b) -27.2 kcal
   c) -191.8 kcal
   d) -135.4 kcal
   e) -59.6 kcal
   f) $+32.6$ kcal
   g) -50.4 kcal
   h) -138.4 kcal
   i) $+57.9$ kcal
   j) -9.9 kcal

13.

	ΔH^0 (kcal)	ΔS^0 (cal deg^{-1})	Direction at high T	Direction at low T
a)	$+12.4$	$+3.3$	Forward	Reverse
b)	-6.6	$+3.2$	Forward	Forward
c)	$+43.2$	$+5.8$	Forward	Reverse
d)	$+9.9$	-10.0	Reverse	Reverse
e)	$+3.1$	$+13.2$	Forward	Reverse
f)	-20.7	$+11.5$	Forward	Forward
g)	$+31.4$	$+31.9$	Forward	Reverse

14. $+4.39$ kcal

15. a) $\Delta H^0 = -17.4$ kcal; $\Delta E^0 = -18.0$ kcal
   b) $\Delta H^0 = -24.8$ kcal; $\Delta E^0 = -24.8$ kcal
   c) $\Delta H^0 = +12.3$ kcal; $\Delta E^0 = +11.7$ kcal
   d) $\Delta H^0 = -68.3$ kcal; $\Delta E^0 = -68.0$ kcal

16. 99.4 kcal mole^{-1}

17. All are available in Appendix 3

18. $D_{C-F} = 117$ kcal mole^{-1}

19. $D_{N-N} = 93.4$ kcal mole^{-1}

Chapter 9

3. a) $\Delta S_{system} = -78.0$ cal deg^{-1}
   b) $\Delta S_{surr} = 458.1$ cal deg^{-1}
   c) $\Delta S_{tot} = 380.1$ cal deg^{-1}

5. a) $\Delta G^0 = -191.4$. Reaction goes to completion.
   b) $\Delta G^0 = -2.2$. Reaction occurs to a limited extent.
   c) $\Delta G^0 = +44.8$. Virtually no reaction occurs.
   d) $\Delta G^0 = +22.5$. Virtually no reaction occurs.
   e) $\Delta G^0 = -16.1$. Reaction goes virtually to completion.
   f) $\Delta G^0 = -123.0$. Reaction goes to completion.
   g) $\Delta G^0 = -8.9$. Reaction occurs to a limited extent.
   h) $\Delta G^0 = -7.8$. Reaction occurs to a limited extent.
   i) $\Delta G^0 = +11.4$. Virtually no reaction occurs.
   j) $\Delta G^0 = +44.1$. Virtually no reaction occurs.
   k) $\Delta G^0 = -9.8$. Reaction goes almost to completion.
   l) $\Delta G^0 = -2.7$. Reaction occurs to a limited extent.

6.

	ΔH^0 (kcal)	$T\Delta S^0$ (kcal)	Direction of reaction determined mainly by:
a)	-191.8	-0.4	ΔH^0
b)	-10.4	-8.2	Both ΔH^0 and $T\Delta S^0$
c)	+36.4	-8.4	ΔH^0
d)	+22.5	0.0	ΔH^0
e)	-25.9	-9.8	ΔH^0
f)	-135.4	-12.3	ΔH^0
g)	-21.0	-12.1	Both ΔH^0 and $T\Delta S^0$
h)	-22.0	-14.1	Both ΔH^0 and $T\Delta S^0$
i)	+12.4	+1.0	ΔH^0
j)	+55.7	+11.5	ΔH^0
k)	-18.4	-8.6	ΔH^0
l)	-11.4	-8.5	Both ΔH^0 and $T\Delta S^0$

Notice that at $25°C$, it is only if ΔH^0 is numerically smaller than about 10 or 12 kcal that ΔS^0 has any influence on the direction of the reaction. In no case is the direction of the reaction determined predominantly by $T\Delta S^0$. These remarks do not, of course, apply at higher temperatures.

8. $289°K = 16°C$

9. $275°K; H_2O$.

10. $950°K$

11. a) From p. 146, $\Delta H^0_{298} = 13.5$ kcal and $\Delta S^0_{298} = 13.1$ cal deg^{-1}. If these are temperature independent, $\Delta G^0 = 0$ at $1030°K$. Below this temperature cyclopentane will be favored, while above it 1-pentene will be favored.

b) From tables, $\Delta H_{298}^0 = -27.2$ kcal and $\Delta S^0 = -30.6$ cal deg^{-1}. These indicate that $\Delta G^0 = 0$ at about $889°K$. Above this temperature reactants will be favored; below it products will be favored.

c) From tables, $\Delta H_{298}^0 = -22.0$ kcal and $\Delta S_{298}^0 = -47.4$ cal deg^{-1}. These values indicate that $\Delta G^0 = 0$ at about $464°K$. Above this temperature N_2 and H_2 will be favored; below it NH_3 will be favored.

d) No NCl_3 is formed at any temperature.

e) $\Delta H^0 \doteq -19$ kcal per ethylene unit from bond energies. $\Delta S^0 \doteq -S^0$ for ethylene $= -52.5$ cal deg^{-1}. From these values, $\Delta G^0 = 0$ at $362°K = 89°C$. Note the low temperature.

f) Diimine decomposes at all temperatures.

12. $\Delta H^0 = 31.4$ kcal; $\Delta S^0 = 31.9$ cal deg^{-1}; $T = 984°K = 711°C$

13.

Gas	ΔH_{298}^0 (kcal)	ΔS_{298}^0 (cal deg^{-1})	$T(\Delta G^0 = 0)$ ($°K$)
CO	257.3	29.1	8842
F_2	37.8	27.4	1380
H_2	104.2	23.6	4420
HI	71.3	21.2	3363
I_2	36.1	24.1	1498
O_2	119.2	28.0	4257

14. $\Delta G_{298}^0 = +72$ kcal;
$T = 2200°K$

15. The valence of all elements would be the same, namely zero.

16. If $\Delta H^0 = -22$ kcal, ΔG^0 will be -10 kcal at $25°C$.

17. The assertion has many exceptions, but it is true that *most* reactions in which more bonds are made than broken are *exothermic,* and also that *most* gaseous reactions which are exothermic at $25°C$ also have a negative value for ΔG^0. (Cf. problems 5 and 6.)

18. $\Delta G^0 = 0$ at approximately $548°K = 275°C$.

Chapter 10

1. $\Delta S = +2.75$ cal deg^{-1}; $\Delta G = -0.821$ kcal.

2. Free energy is not really an energy and is not conserved. To say that it "gets lost" implies that it can be recovered.

3. Yes! Work *must* be done in order to compress it.

4. 4/27; 1/27.

5. No change in entropy occurs (no change in *state* occurs).

6. $\Delta S = +2.99$ cal deg^{-1}.

7. $S = 254.4$ cal deg^{-1}.

8.

T	ΔS (cal deg^{-1})	ΔH (kcal)	ΔG (kcal)
$298°$K	3.79	0.0	-1.13
$500°$K	3.79	0.0	-1.90

12. $S = 1/3 R \log_e (3/1) + 2/3 R \log_e (3/2)$

Chapter 11

1. a) $K_p = (P_{HI})^2 / P_{H_2} \times P_{I_2}$
b) $K_p = (P_{HI})^2 / P_{H_2}$
c) $K_p = P_{H_2O} / P_{H_2}$
d) $K_p = P_{CS_2}$
e) $K_p = P_{CO_2}$

2.

ΔG^0 (kcal)	Extent of reaction
a) $+53.1$	0.000
b) -2.6	0.96
c) -1.6	0.93
d) -236.6	1.000
e) $+41.4$	0.000
f) -2.2	0.7
g) $+46.8$	0.000
h) $+0.3$	1.000 (Think!)
i) $+34.9$	0.000
j) $+8.4$	0.000
k) -2.7	0.8
l) -31.6	1.000
m) -37.6	1.000

3. $\Delta H^0 = 0$, $\Delta S^0 = 0$, $\Delta G^0 = 0$, $K = 1$

4. 119 mm (CCl_4); 390 mm (CS_2)

5. $77°$C (CCl_4); $47°$C (CS_2)

6. Below $458°$K, I_2 is a liquid at atmospheric pressure, and below $387°$K it is a solid.

8. $\Delta G^0 = +6.15$ kcal (from K_p)
$\Delta G^0 = +5.68$ kcal (from tables)

9. $K_p = 2.56 \times 10^{-2}$ (from experimental data)
  $= 2.38 \times 10^{-2}$ (from ΔG_f^0 tables)

10. $\Delta G^0 = -5.56$ kcal (from experimental K_p)
  $= -5.61$ kcal (from tables)

11. $K_p = 4.45 \times 10^{-2}$ (from experimental data)
  $= 3.35 \times 10^{-2}$ (from tables)

12. 41.6% isopentane and 58.3% n-pentane.

13. 1.44 moles at 1 atmosphere. 0.63 mole at 10 atmospheres.

14. 0.0838 mole NH_3 calculated from tables as compared to 0.0797 mole NH_3 observed experimentally.

15. Use of the approximate value of ΔG^0 (9.91 kcal) yields the prediction that 0.0312 mole of NH_3 will be formed.

16. $\Delta G_{318}^0 = +0.29$ kcal; $K_p = 0.632$. The calculated result is that 0.610 mole of N_2O_4 will dissociate.

17. $\Delta G_{502}^0 = +0.56$ kcal; $K_p = 0.571$. The calculated result is that 0.607 mole of PCl_5 will dissociate.

18. $\Delta H^0 \doteq -41$ kcal from bond energies; $\Delta S^0 \doteq -30$ cal deg^{-1}, a little smaller than $S^0(H_2)$. This gives the result that $\Delta G = 0$ at $1370°K$, and that $\Delta G_{298}^0 = -32$ kcal.

19. $\Delta G_{298}^0 = -33.7$ kcal

20. $943°K$

21. $\Delta G_{500}^0 = +0.84$ kcal

22. If neither ΔH^0 nor ΔS^0 varies with T, then ΔG^0 will vary in a linear manner with T because of the relationship $\Delta G^0 = \Delta H^0 - T\Delta S^0$. When ΔG^0 is plotted against T, the slope is $-\Delta S^0$ and the intercept at $T = 0°K$ is ΔH^0. From such a graph we find $\Delta H^0 = 41.2$ kcal and $\Delta S^0 = 35.1$ cal deg^{-1}. The values for $25°C$ obtained from Appendix 3 are $\Delta H^0 = 42.4$ kcal and $\Delta S^0 = 38.4$ cal deg^{-1}.

23. $\Delta G = -411$ cal. Free energy is not an energy and is not conserved. Free energy cannot "get lost" since this implies we can recover it again later.

24. ΔG^0 for the reaction

$$A(l) \rightarrow A(g)$$

is related to both ΔH^0 and ΔS^0 on the one hand and the vapor pressure P on the other hand.

$$\Delta G^0 = -RT \log_e P$$

$$= \Delta H^0 - T\Delta S^0$$

If ΔH^0 and ΔS^0 do not vary with T we have

$$\log_e P = \frac{-\Delta H^0}{RT} + \Delta S^0$$

so that $A = -\Delta H^0/R$ while $B = \Delta S^0$

26. At $500^\circ K$, $\Delta G^0 = +4.90$ kcal and the extent is 0.078.
    At $1000^\circ K$, $\Delta G^0 = -0.73$ kcal and the extent is 0.546.

INDEX